Lecture Notes in Morphogenesis

Series editor

Alessandro Sarti, CAMS Center for Mathematics, CNRS-EHESS, Paris, France
e-mail: alessandro.sarti@ehess.fr

More information about this series at http://www.springer.com/series/11247

Mirko Degli Esposti · Eduardo G. Altmann
François Pachet

Editors

Creativity and Universality in Language

 Springer

Editors
Mirko Degli Esposti
Dipartimento di Matematica
Università di Bologna
Bologna
Italy

François Pachet
Sony Computer Science Laboratory
Paris
France

Eduardo G. Altmann
Max Planck Institute for the Physics
  of Complex Systems
Dresden
Germany

ISSN 2195-1934   ISSN 2195-1942   (electronic)
Lecture Notes in Morphogenesis
ISBN 978-3-319-79621-5   ISBN 978-3-319-24403-7   (eBook)
DOI 10.1007/978-3-319-24403-7

© Springer International Publishing Switzerland 2016
Softcover reprint of the hardcover 1st edition 2016
This work is subject to copyright. All rights are reserved by the Publisher, whether the whole or part
of the material is concerned, specifically the rights of translation, reprinting, reuse of illustrations,
recitation, broadcasting, reproduction on microfilms or in any other physical way, and transmission
or information storage and retrieval, electronic adaptation, computer software, or by similar or dissimilar
methodology now known or hereafter developed.
The use of general descriptive names, registered names, trademarks, service marks, etc. in this
publication does not imply, even in the absence of a specific statement, that such names are exempt from
the relevant protective laws and regulations and therefore free for general use.
The publisher, the authors and the editors are safe to assume that the advice and information in this
book are believed to be true and accurate at the date of publication. Neither the publisher nor the
authors or the editors give a warranty, express or implied, with respect to the material contained herein or
for any errors or omissions that may have been made.

Printed on acid-free paper

This Springer imprint is published by Springer Nature
The registered company is Springer International Publishing AG Switzerland

Preface

How can we characterize originality and innovation in authors or texts? How can we reveal universal features of language? The aim of this volume is to address these questions by confronting different quantitative approaches to originality and universality in language. New methods have shown to be increasingly successful in addressing the traditional problems of authorship attribution and document classification. These results provide insights on how to quantify the unique features of authors, composers, and styles. Such features contrast, and are restricted by, universal properties of texts, such as scaling laws in word-frequency distribution, entropy measures, and long-range correlations. This interplay between innovation and universality is also an essential ingredient of methods for automatic text generation and of models of linguistic innovations. This volume collects contributions from scientists with different backgrounds interested in quantitative analysis of variations (synchronic and diachronic) in language. The aim is to obtain a deeper understanding of how originality emerges, can be quantified, and propagates.

Bologna   Mirko Degli Esposti
Dresden   Eduardo G. Altmann
Paris   François Pachet
January 2016

Acknowledgments

This book originates in a workshop held in June 2014 in Paris and entitled *Creativity and Universality in Language*.[1] This workshop was organized within the ERC project *Flow Machines*,[2] which addresses the issue of modeling style in music and text. The Flow Machines project received funding from the European Research Council under the European Union Seventh Framework Programme (FP/2007–2013)/ERC Grant Agreement n. 291156.

We thank Jean-Francois Perrot for his contribution to the workshop, and friendly support throughout the editing process. We also thank Fiammetta Ghedini for her decisive contributions to the dissemination of the project, and Giampaolo Cristadoro and Pierre Roy for their help in the organization of the workshop.

---

[1]http://www.flow-machines.com/textandcreativityworkshop2014.
[2]http://flow-machines.com.

Acknowledgments

Contents

Contributors

Eduardo G. Altmann Max Planck Institute for the Physics of Complex Systems, Dresden, Germany

Dario Benedetto Dipartimento di Matematica, Sapienza Università di Roma, Rome, Italy

Richard A. Blythe School of Physics and Astronomy, SUPA, University of Edinburgh, Edinburgh, UK

Fabio Celli University of Trento, Trento, Italy

Mirko Degli Esposti Dipartimento di Matematica, Università di Bologna, Bologna, Italy

Martin Gerlach Max Planck Institute for the Physics of Complex Systems, Dresden, Germany

Pablo Gervás Instituto de Tecnología del Conocimiento, Universidad Complutense de Madrid, Madrid, Spain

Delia Irazú Hernández Farías Natural Language Engineering Lab, Universitat Politècnica de València, Valencia, Spain

Cyril Labbé LIG, University of Grenoble Alpes, Grenoble, France; LIG, CNRS, Grenoble, France

Dominique Labbé PACTE, University of Grenoble Alpes, Grenoble, France; PACTE, CNRS, Grenoble, France

Carlos León Departamento de Ingeniería del Software e Inteligencia Artificial, Universidad Complutense de Madrid, Madrid, Spain

Vittorio Loreto Physics Department, Sapienza University of Rome, Rome, Italy; ISI Foundation, Torino, Italy; SONY-CSL, Paris, France

Marcelo A. Montemurro Faculty of Life Sciences, The University of Manchester, Manchester, UK

François Pachet Sony CSL, Paris, France

Alexandre Papadopoulos UPMC Paris 6, UMR 7606, LIP6, Paris, France

François Portet LIG, University of Grenoble Alpes, Grenoble, France; LIG, CNRS, Grenoble, France

Francisco Rangel Natural Language Engineering Lab, Universitat Politècnica de València, Valencia, Spain; Autoritas Consulting S.A., Valencia, Spain

Paolo Rosso Natural Language Engineering Lab, Universitat Politècnica de València, Valencia, Spain

Pierre Roy Sony CSL, Paris, France

Vito D.P. Servedio Physics Department, Sapienza University of Rome, Rome, Italy; Institute for Complex Systems (CNR-ISC), Rome, Italy

Efstathios Stamatatos University of the Aegean, Karlovassi, Greece

Luc Steels Institute for Advanced Studies (ICREA), IBE (UPF-CSIC), Barcelona, USA

Steven H. Strogatz Cornell University, Ithaca, NY, USA

Francesca Tria ISI Foundation, Torino, Italy

Damián H. Zanette Centro Atómico Bariloche e Insituto Balseiro, San Carlos de Bariloche, Río Negro, Argentina

Introduction to the Volume

Mirko Degli Esposti, Eduardo G. Altmann and François Pachet

1 Why This Volume?

Like most human productions, language is the product of cultural evolution, and as such exhibits high levels of complexity. A natural representation of language is written text, an expression of language by letters or other marks. Preceded by proto-writing systems of ideographic and/or early mnemonic symbols, so-called *true writing*, in which the content of a linguistic utterance is encoded so that another reader can reconstruct, with a fair degree of accuracy, the exact utterance written down, characterizes human evolution since 3200 BC. Development of writing coincides with the development of literature and both phenomena have been highly effected by the development of technologies, up to the enormous proliferation of text (at least quantitatively) in our modern digitized societies. It is commonly acknowledged that the main purpose of written texts is to carry, transmit, and preserve meaning, or information. As such, text writing and reading have been central subjects of human studies throughout history.

However, texts can also be studied independently of their meaning or interpretation, as natural artifacts produced by our society. In the past few years, an increasing number of scientists have been interested in studying text from such an agnostic, physics perspective, and this book attempts to capture recent trends in this direction. More precisely, we address here questions such as how to characterize *originality* in

M. Degli Esposti (✉)
Dipartimento di Matematica, Università di Bologna, Bologna, Italy
e-mail: mirko.degliesposti@unibo.it

E.G. Altmann
Max Planck Institute for the Physics of Complex Systems, Dresden, Germany
e-mail: edugalt@pks.mpg.de

F. Pachet
Sony CSL, 6 rue Amyot, 75005 Paris, France
e-mail: pachetcsl@gmail.com

© Springer International Publishing Switzerland 2016
M. Degli Esposti et al. (eds.), *Creativity and Universality in Language*,
Lecture Notes in Morphogenesis, DOI 10.1007/978-3-319-24403-7_1

authors or texts, and how innovations originate amid universal features of language usage?

The contributions in this book connect text analysis and text generation by focusing on the interplay between invariant (*universal*) and unique (*creative*) features of language use. Universal properties of texts are interpreted here as boundaries that regulate the creative process of language production, bringing new insights about the natural production of language and, more practically, about the constraints that need be imposed for more efficient, reliable and convincing automatic text generation. The most interesting properties of specific texts are often their nonuniversal properties, which can be exploited in quantitative stylometry, authorship attribution and document classification methods, to mention a few applications. Language usage is of course a highly nonstationary process, affected by cultural, social and technological factors. Universality of the process of text production combined with creativity (in language innovations) lie at the heart of language changes and evolution. The systematic digitization of human communication brings new scientific challenges both to automated text analysis and automatic text generation, and motivate a combined view on both issues as addressed in this book.

Understanding the fundamental processes and laws underlying creativity and innovation in generating textual data, measuring and simulating different writing styles, understanding the evolution in time of languages and literary styles, exploring novel quantitative methods to detect plagiarism and to automatically reveal fake or artificial documents, such as scientific papers, are all different sides of a common research objective that involves several, sometimes distant disciplines, such as mathematics, computer science, physics, and linguistics. The complexity of many of the considered problems originates from the fact that (written) language is a multi-layer system, that can be represented at different but correlated scales, e.g., phonetical, grammatical syntactical, and semantical. While language still remains out of reach for the most modern artificial intelligence (AI) systems, multi-scale approaches are necessary, in particular for solving practical tasks such as automatic translation, semantic extraction, authorship attribution,and automatic summarization. The contributions in this volume as well as their order reflect such a multi-scale nature of language and textual data in particular.

2 Contributions

The first two contributions focus on linguistic laws emerging at the level of words usage in natural languages.

(1) The first contribution by Eduardo G. Altmann and Martin Gerlach offers a critical look at the statistical interpretations of linguistic laws. The famous *Zipf's law*, which relates the frequency f of the rth most frequent word to its rank ($f \propto 1/r$), is only one out of many laws proposed to describe statistical regularities in language. The availability of large databases of written data allows the authors to perform deep statistical analysis and a rigorous exploration of the fluctuations around the proposed law. The presence of correlations and fluctuations are critically discussed and alternative interpretations of the results are presented, suggesting that proper statistical analysis should include null models that account for the observed fluctuations and correlations. These results about linguistic laws have both theoretical and practical impact.

(2) The second contribution by Marcelo A. Montemurro and Damian H. Zanette focuses on statistical properties of word usage in different languages, with an emphasis on the statistical structures emerging as a balance between order (phonetical, grammatical, and syntactical constraints) and disorder (semantical choices and author's style). This balance between order and disorder is necessary to enable a high information rate and at the same time robustness under communications errors. As discussed in this contribution, these constraints contribute to shape the statistical structure of linguistic sequences. A measure of relative entropy that quantifies the degree of order in word patterns is reviewed, discussed, and applied on language samples of 24 linguistic families to assess the contribution of word ordering to the statistical structure of language. While, during evolution, different languages have developed different vocabularies and sets of rules, the data suggest that the evolution and the related diversification among languages have been constrained to an almost constant value of the relative entropy. Another entropy measure, related to the spatial distribution of words in a given text, is able to characterize the semantic role of words and automatically extract from a given text the words that are most closely related to their semantic content and also to highlight semantic relationships between them.

The third and fourth contribution address the mechanisms underlying language evolution and language innovation.

(3) Richard A. Blythe explores universal mechanisms in language change (or cultural evolution, more generally) where an innovation (a new word for example) is introduced and then spread through social interactions to replace the existing convention. When one looks at these phenomena arising from repeated social interactions, common patterns emerge in an *universal* way, in the sense that they have been observed in different social groups at different times and also across several cultural behaviors. Using explicit examples, the author argues that such universal patterns of cultural evolution are in fact related to the universal constraints underlying social interactions and social learning, determined by the cognitive and physical apparatus possessed by the interacting agents.

(4) The contribution by Vittorio Loreto, Vito D. P. Servedio, Steven H. Strogatz, and Francesca Tria offers a detailed and complete overview of scientific attempts to model the emergence of novelties in language (Heaps' law) as well as in other social, biological, and technological systems: from the plain Simon model of the 1950s, to the latest model of Polya's urn with triggering. As discussed in this chapter, we experience novelties very often in our daily lives. We meet new people, adopt new words, listen to new songs, watch a new movie, use a new technology. This is a very significant phenomenon often referred to as innovation, a fundamental factor in the evolution of biological systems, human society, and technology. From this

perspective, a thorough investigation and a deep understanding of the underlying mechanisms through which novelties and innovations emerge, diffuse, compete, and stabilize is a key to progress in all sectors of human activities. They introduce an original mathematical model of the dynamics of novelties, from which they derive three testable, quantitative predictions. The model allows them to explain the statistical laws of the rate at which novelties happen (Heaps' law) and of the frequency distribution of the explored regions of the space (Zipf's law), as well as the signatures of the correlation process by which one novelty sets the stage for another. The predictions of this models were tested on four data sets of human activity: the edit events of Wikipedia pages, the emergence of tags in social annotation systems, the sequence of words in texts, and listening to new songs in on-line music catalogues.

The next three contributions address methods for automatic generation of texts in a given style.

(5) The contribution by Alexandre Papadopoulos, François Pachet, and Pierre Roy explore Markov chains to analyze and generate sequences that imitate a given style. Combining techniques from constraint satisfaction, automata theory, and statistical inference they generate *novel* sequences that "look" like or "sound" like the original ones, with a guarantee of non-plagiarisms. Technically, they address the generation of Markov sequences with a guaranteed *maximum order*. More precisely, they address the problem of Markov sequence generation with forbidden k-gram constraints. This problem is addressed in two steps. In the first step, they show that, given a Markov transition matrix and a set of k-grams, one can build efficiently an automaton that represents exactly the language of all sequences that can be generated (efficiently in both space and time) from a Markov model, and that also do not contain any of the k-grams. In the second step, they show that the automaton can be extended so as to be exploited by a belief propagation scheme, in order to produce unbiased sampling of all the solutions.

(6) Pablo Gervás and Carlos León review recent advances in the area of computational creativity devoted to generative processes and evaluation models of literary artifacts (e.g., poetry generation and story telling), a long-standing dream of artificial intelligence. As the authors explain, in the early times of AI these approaches very rarely modeled the iterative nature of creative process as observed in humans, where a creator sets out with a purpose in mind, and creates drafts and revises them successively until the purpose is met. While it is a fact that in truly creative processes the purpose may also evolve during revision, the lack of direction has always been a damaging criticism to early approaches to computational creativity. For the authors this shows the necessity of introducing an evaluation process to drive the construction. In their contribution they address a set of examples of approaches to the computational generation of literary texts based on particular techniques, and describe a computational model of the creative process for literary texts (the ICTIVS model) that captures the purpose-driven revision of generated artifacts.

(7) Cyril Labbé, Dominique Labbé, and François Portet discuss the detection of computer generated papers in scientific literature, a problem of increasing practical importance. They start by reviewing how meaningless computer generated scientific

texts allowed *Ike Antkare* to become one of the most highly cited scientists of the modern world. But such fake publications also appear in real scientific conferences and, as a result, in bibliographic databases. Recently, more than 120 papers have been withdrawn from subscription databases of two high-profile publishers, IEEE and Springer, because they were computer generated with SCIgen. SCIgen is a software based on a probabilistic context free grammar (PCFG) designed to randomly generate computer science research papers. In this contribution, they first describe two different types of natural language generation, Markov chains and probabilistic context free grammar, emphasizing their main lexical and stylistic differences.

Related to the automatic generation of texts under constraints, the next two contributions discuss the problem of authorship attribution.
(8) Efstathios Stamatatos reviews and discusses the different facets of what we call the *style* of a document and the great number of measures proposed to quantify writing style. He detects features that can be characterized as universal, in the sense that they can be easily extracted from any kind of text in practically any natural language. He describes the basic categories of stylometric features found in style-based text categorization tasks, mainly authorship attribution and genre identification. He examines two types of such universal stylometric features: function words and character n-grams. These can easily be extracted from any document and have been successfully used in several style-based text categorization tasks, like authorship attribution, automatic genre identification, and plagiarism detection. Focusing on authorship attribution, the author explores the effectiveness of these features in difficult cases where there are differences in topic and genre of the documents under examination. It is demonstrated that one crucial decision concerns the representation of dimensionality and that it is possible to reach high accuracy results provided that the appropriate number of features is used.
(9) Focusing on authorship attribution, Dario Benedetto and Mirko Degli Esposti introduce and discuss how quantitative methods based on fundamental principles of information theory can be used to address a concrete problem of attribution. They describe and discuss a recent dispute concerning the authenticity of the *Diario Postumo*, a collection of 84 poems hitherto attributed to Eugenio Montale, an Italian poet, laureate of the 1975 Nobel Prize in Literature. In such a real, and peculiar, case, far from artificially designed experiments, these mathematical methods do not allow for a definitive and scientific solid conclusion. However, they can be efficiently coupled with historical and philological considerations to form an effective multidisciplinary approach to this attribution problem, revealing new shadows and doubts on the integrity and on the authenticity of the *Diario Postumo*.

Authorship attribution is quite a specific task that can be generalized to the more abstract task of author profiling, i.e., the characterization of personal and social aspects of the authors of textual material, such as gender, age, native language, personality type, social position, geographical information, etc. The following two contributions focus on these topics.
(10) Paolo Rosso, Irazú Hernández Farias and Francisco Rangel explore author profiling, dealing with distinguishing between classes of authors rather than individual

authors on the basis of their usage of language. In particular, they focus on the use and on the detection of irony, a subjective usage of language, as a linguistic structure to transmit information. Quoting the authors, the aim of this chapter is to introduce the reader to concepts such as universality of language among classes of authors, e.g., of the same gender, and creativity in irony. After an exhaustive review of author profiling tasks and recent methods, the authors investigate the impact of emotions on author profiling, using a specific style-based and emotion-labeled graph (Emo-Graph), built from textual features. Starting from what was already investigated for English, they explore the use of the different morphosyntactic categories in Spanish. They compare different approaches showing how best results are achieved with Support Vector Machines on a specific EmoGraph approach (Rangel-EG).

(11) In his contribution, Fabio Celli addresses the issue of computational human creativity analysis (HCA) from a natural language processing (NLP) perspective. He introduces a computational framework for creativity analysis with two approaches: an agnostic one, based on clustering, and a knowlegde-based one, that exploits supervised learning and feature selection. The contribution starts with a brief but effective survey of how the issue of creativity has been addressed in psychology and the humanities, shifting then to the description of computational approaches to human creativity. In particular the author devotes his attention to artificial creativity, the study of the creative behaviour of individuals in societies by means of networks of agents, whose parameters and interactions can be observed in a controlled setting. The author then defines a framework for exploiting NLP techniques for human creativity analysis, focusing on styles and their relationships with creativity.

(12) Finally, Luc Steels examines two issues which have not yet been adequately handled by statistical/probabilistic approaches to language processing: meaning and creativity. The author argues that meaning can be tackled by shifting from corpus-based learning to learning through language-games. As analyzed in the contribution, corpora usually contain only utterances without a representation or enough contextual information of the communicative goals and semantic structures that underlie these utterances. On the other hand, language games introduce meaning because the speaker and listener have shared goals and a shared context, enabling the generation of meaning. The author then goes a step further by arguing that creativity can be brought in by introducing a two-tier architecture with routine language processing interlaced with meta-level insight problem solving, to handle the intentional deviations that speakers introduce in order to adapt and expand their language systems.

Enjoy the reading!

Statistical Laws in Linguistics

Eduardo G. Altmann and Martin Gerlach

Abstract Zipf's law is just one out of many universal laws proposed to describe statistical regularities in language. Here we review and critically discuss how these laws can be statistically interpreted, fitted, and tested (falsified). The modern availability of large databases of written text allows for tests with an unprecedent statistical accuracy and also for a characterization of the fluctuations around the typical behavior. We find that fluctuations are usually much larger than expected based on simplifying statistical assumptions (e.g., independence and lack of correlations between observations). These simplifications appear also in usual statistical tests so that the large fluctuations can be erroneously interpreted as a falsification of the law. Instead, here we argue that linguistic laws are only meaningful (falsifiable) if accompanied by a model for which the fluctuations can be computed (e.g., a generative model of the text). The large fluctuations we report show that the constraints imposed by linguistic laws on the creativity process of text generation are not as tight as one could expect.

> "... 'language in use' cannot be studied without statistics" Gustav Herdan (1964) [1]

1 Introduction

In the past 100 years regularities in the frequency of text constituents have been summarized in the form of *linguistic laws*. For instance, Zipf's law states that the frequency f of the rth most frequent word in a text is inversely proportional to its rank: $f \propto 1/r$ [2]. This and other less famous linguistic laws are one of the main objects of study of *quantitative linguistics* [3–8].

Linguistic laws have both theoretical and practical importance. They provide insights on the mechanisms of text (language, thought) production and are also crucial in applications of statistical natural language processing (e.g., information retrieval).

E.G. Altmann (✉) · M. Gerlach
Max Planck Institute for the Physics of Complex Systems, Dresden, Germany
e-mail: edugalt@pks.mpg.de

M. Gerlach
e-mail: gerlach@pks.mpg.de

© Springer International Publishing Switzerland 2016
M. Degli Esposti et al. (eds.), *Creativity and Universality in Language*,
Lecture Notes in Morphogenesis, DOI 10.1007/978-3-319-24403-7_2

Both the generative and data analyses views of linguistic laws are increasingly important in modern applications. Data mining algorithms profit from accurate estimations of the vocabulary size of a collection of texts (corpus), e.g., through Heaps' law discussed in the next section. Methods for the automatic generation of natural language can profit from knowing the linguistic laws underlying usual texts. For instance, linguistic laws may be included as (additional) constraints in the space of possible (Markov generated) texts [9] and can thus be considered as constraints to the creativity of authors.

Besides giving an overview on various examples of linguistic laws (Sect. 2), in this chapter we focus on their probabilistic interpretation (Sect. 3), we discuss different statistical methods of data analysis (Sect. 4), and the possibilities of connecting different laws (Sect. 5). The modern availability of large text databases allows for an improved view on linguistic laws that requires a careful discussion of their interpretation. Typically, more data confirms the observations motivating the laws—mostly based on visual inspection—but makes it increasingly difficult for the laws to pass statistical tests designed to evaluate their validity. This leads to a seemingly contradictory situation: while the laws allow for an estimation of the general behavior (e.g., they are much better than alternative descriptions), they are strictly speaking falsified. The aim of this contribution is to present this problem and discuss alternative interpretations of the results. We argue that the statistical analysis of texts often shows long-range correlations and large (topical) fluctuations. We conclude that proper statistical analysis of linguistic laws, including tests of their validity, should consider (null) models that account for the observed fluctuations and correlations.

2 Examples and Observations

An insightful introduction to Linguistic Laws is given in Ref. [5] by Köhler, who distinguishes between three kinds of laws as follows:

1. "The first kind takes the form of probability distributions, i.e., it makes predictions about the number of units of a given property."
2. "The second kind of law is called the functional type, because these laws link two (or more) variables, i.e., properties."
3. "The third kind of law is the developmental one. Here, a property is related to time." (time may be measured in terms of text length)

We use the term linguistic law to denote quantitative relationships between measurements obtained in a written text or corpus, in contrast to syntactic rules and to phonetic and language change laws (e.g., Grimm's law, see also the chapter by R. Blythe in this book). We assume that the laws make statements about individual

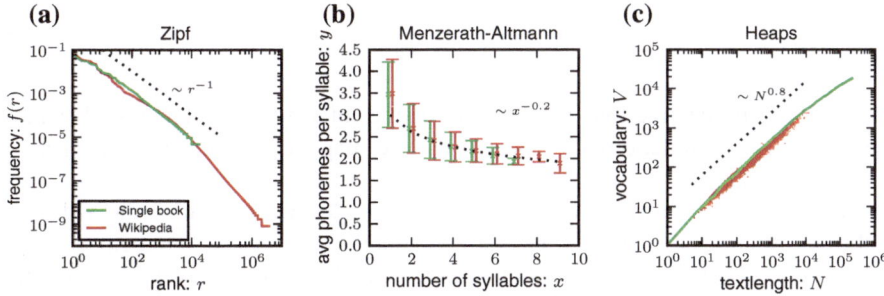

Fig. 1 Examples of linguistic laws: **a** Zipf, **b** Menzerath–Altmann, and **c** Heaps' laws. Data from one book (*green*, Moby Dick by H. Melville) and for the English Wikipedia (*red*) are shown. *Dotted* (*black*) *lines* are the linguistic laws with arbitrary parameter, chosen for visual comparison (see Appendix for details)

texts (corpus) and are exact in an appropriate limit (e.g., large corpus).[1] Each law contains parameters which we denote by Greek letters α, β, γ, and often refer to the frequency $f(q)$ of a quantity q in the text (with $\sum_q f(q) = 1$). Probabilities are denoted by $P(q)$.

Next we discuss in detail one representative example of each of the three types of laws mentioned above: Zipf, Menzerath–Altmann, and Heaps' laws, respectively, see Fig. 1.

1. Zipf's law is the best-known linguistic law (see, e.g., Ref. [10] for historical references). In an early and simple formulation, it states that if words (types) are ranked according to their frequency of appearance $r = 1, 2, \ldots, V$, the frequency $f(r)$ of the rth word (type) scales with the rank as

$$f(r) = \frac{f(1)}{r}, \tag{1}$$

where $f(1)$ is the frequency of the most frequent word. The above expression cannot hold for large r because for any $f(1) > 0$, there exist an r^* such that $\sum_{r=1}^{r^*} f(1)/r > 1$. Taking also into account that $f(1)$ may not be the best proportionality factor, a modern version of Zipf's law is

$$f(r) = \frac{\beta_z}{r^{\alpha_z}}, \tag{2}$$

with $\alpha_z \gtrsim 1$, see Fig. 1a. The analogy with other processes showing fat-tailed distribution motivates the alternative formulation

---

[1] While some of the laws clearly intend to speak about the language as a whole, in practice they are tested and motivated by observations in specific texts which are thus implicitly or explicitly assumed to reflect the language as a whole.

$$P(f) = \frac{\beta_Z^\dagger}{f^{\alpha_Z^\dagger}}, \tag{3}$$

where $P(f)$ is the fraction (probability) of the total number of words that have frequency f. Formulations (2) and (3) can be mapped to each other with $\alpha^\dagger = 1 + 1/\alpha$ [10–12].

2. Menzerath–Altmann law received considerable attention after the works of Gabriel Altmann [4–6, 13]. Menzerath's general (qualitative) statement originating from his observations about phonemes is that "the greater the whole the smaller its parts." The quantitative law intended to describe this observation is [13]

$$y = \alpha_M x^{\beta_M} e^{-\gamma_M x}, \tag{4}$$

where x measures the length of the whole and y the (average) size of the parts. One example [13] is obtained computing for each word w the number of syllables x_w and the number of phonemes z_w. The length of the word (the whole) is measured by the number of syllables x_w, while the length of the parts is measured for each word as the average number of phonemes per syllable $y_w = z_w/x_w$. The comparison to the law is made by averaging y_w over all words w with $x_w = x$, see Fig. 1b. The ideas of Menzerath–Altmann law and Eq. (4) have been extended and applied to a variety of problems, see Ref. [14] and references therein.

3. Heaps' law states that the number of different words V (i.e., word types) scales with database size N measured in the total number of words (i.e., word tokens) as [1, 15]

$$V \sim N^{\alpha_H}. \tag{5}$$

In Fig. 1c this relationship is shown in two different representations. For a single book, the value of N is increased from the first word (token) until the end of the book so that $V(N)$ draws a curve. For the English Wikipedia, each article is considered as a separate document for which V and N are computed and shown as dots.

The nontrivial regularities and the similarity between the two disparate databases found for the three cases analyzed in Fig. 1 strongly suggest that the three linguistic laws summarized above capture important properties of the structure of texts. Additional examples of linguistic laws are listed in Table 1, see also the vast literature in quantitative linguistics [3–6]. The (qualitative) observations reported above motivate us to search for quantitative analysis that match the requirements of applications and the accuracy made possible through the use of large corpora. The natural questions that we would like to address here are: Are these laws true (compatible with the observations)? How to determine their parameters? How much fluctuations around them should be expected (allowed)? Are these laws related to each other? Before addressing these questions we discuss *how should one interpret linguistic laws.*

Table 1 List of linguistic laws

Name of the law	Observables	Functional form	References
Zipf	f: freq. of word w; r: rank of w in f	$f(r) = \beta_Z r^{-\alpha_Z}$	[2, 10, 11, 16–19]
Menzerath–Altmann	x : length of the whole; y : size of the parts	$y = \alpha_M x^{\beta_M} e^{-\gamma_M x}$	[13, 14]
Heaps	V : number of words; N : database size	$V \sim N^{\alpha_H}$	[1, 15, 20–24]
Recurrence	τ : distance between words	$P(\tau) \sim \exp(\alpha\tau)^{\beta}$	[2, 25–27]
Long-range correlation	$C(\tau)$: autocorrelation at lag τ	$C(\tau) \sim \tau^{-\alpha}$	[28–30]
Entropy scaling	H : entropy of text with blocks of size n	$H \sim \alpha n^{\beta} + \gamma n$	[31, 32]
Information content	$I(l)$: information of word with length l	$I(l) = \alpha + \beta l$	[2, 33]
Taylor's law	σ: standard deviation around the mean μ	$\sigma \sim \mu^{\alpha}$	[24]
Networks	Topology of lexical/semantic networks	Various	[34–37]

3 Interpretation of Linguistic Laws

In Chap. 26 *Text Laws* of Ref. [3], Hřebiček argues that

> ...the notion *law* (in the narrower sense *scientific law*) in linguistics and especially in quantitative linguistics ... need not obtain some special comprehension different from its validity in other sciences. Probably, the best delimitation of this concept can be found in the works by the philosopher of scientific knowledge Karl Raimund Popper...

This view is also emphasized by Köhler in Ref. [5], who distinguishes *laws* from *rules* and states that a *"significant difference is that rules can be violated—laws (in the scientific sense) cannot."*.

Such a straightforward identification between linguistic and scientific laws masks the central role played by statistics (and probability theory) in the interpretation of linguistic laws. To see this, first notice that these laws do not directly affect the production of (grammatically and semantically) meaningful sentences, e.g., because they involve scales much larger or shorter than a sentence. It is thus not difficult to be convinced that a creative and persistent daemon,[2] trained in the techniques of *constrained writing* [38] (see also the chapter by A. Papadopoulos, F. Pachet, and P. Roy in this book), can generate understandable and arbitrary long texts which deliberately violate any single law mentioned above. In a strict Popperian sense, a single

---

[2]A relative of Maxwell's Daemon known from Thermodynamics.

of such demonic texts would be sufficient to falsify the proposed laws. Linguistic laws are thus different from syntactic rules and require a different interpretation than, e.g., the laws of classical physics.

The central role of statistics in Quantitative Linguistics was emphasized by its founding father Gustav Herdan:

> The distinction between language laws in the conventional sense and statistical laws of language corresponds closely to that between the classical laws of nature, or the physical universe, and the statistical laws of modern physics [1].

Altmann, when discussing Menzerath law [13], also emphasizes that "this law is a stochastic one," and Köhler [3] refers to the concept of stochastic hypothesis. There are at least two instances in which a statistical interpretation should be included:

1. In the statement of the law, e.g., in Zipf's law (3) the probability of finding a word with frequency f decays as $P(f) \sim f^{-\alpha_z^\dagger}$.
2. In the interpretation of the law as being *typical* in a collection of texts, e.g., in Heaps' law the vocabulary V of a typical text of size N is $V \sim N^{\alpha_H}$.

The demonic texts mentioned above would be considered *untypical* (or highly unlikely). Statistical laws in at least one of these senses are characteristic not only of modern physics, as pointed out by Herdan, but also of different areas of natural and social sciences: Benford's law predicts the frequency of the first digit of numbers appearing in a corpus [39] and the Gutenberg–Richter law determines the frequency of earthquakes of a given magnitude [40]. The analysis of these laws, including possible refutations, have to be done through statistical methods, the subject of the next section. Important aspects of linguistic laws not discussed in detail in this chapter include: (i) the universality and variability of parameters of linguistic laws (e.g., across different languages [21, 22, 37, 41, 42] (see also the chapter by M.A. Montemurro and D.H. Zannette in this book), as a function of size [43] and degree of mixture of the corpus [44], styles [29], and age of speakers [45]); and (ii) the relevance and origins of the laws. This second point was intensively debated for Zipf's law [8, 19, 46], with quantitative approaches based on stochastic processes—e.g., the Monkey typewriter model [11, 47] and rich-get-richer mechanisms [10, 11, 16, 18, 22]—and on optimization principles—e.g., between speaker and hearer [2, 48], of the mutual information between forms and meanings [49], or of general entropy maximization principles [50, 51].

4 Statistical Analysis

In Sect. 2 we argued in favor of linguistic laws by showing a graphical representation of the data (Fig. 1). The widespread availability of large databases and the applications of linguistic laws require and allow for a more rigorous statistical analysis of the results. To this end we assume the linguistic law can be translated into a precise mathematical statement about a curve or distribution, which contain a set of parameters. Legitimate questions to be addressed are:

(1) **Fitting**. What are the best parameters of the law to describe a given data?
(2) **Model Comparison**. Is the law better than an alternative one?
(3) **Validity**. Is the law compatible with the observations?

These points are representative of statistical analysis performed more generally and should preceed any more fundamental discussion on the origin and importance of a specific law. Below we discuss in more details how each of the three points listed above has been and can be addressed in the case of linguistic laws.

4.1 Graphical Approaches

Visual inspection and graphical approaches were the first type of analysis of linguistic laws and are still widely used. One simple and still very popular fitting approach is least squares (minimize the squared distance between data and models). Often this is done in combination with a transformation of variables that maps the law into a straight line (e.g., using logarithmic scales in the axis or taking the logarithm of the independent and dependent variable in the Zipf's and Heaps' laws). These transformations are important to visually detect patterns and are parts of any data analysis. However, they are not appropriate for a quantitative analysis of the data. The problem of fitting straight lines in log–log scale is that least square fitting assumes an uncertainty (fluctuation) on each point that is independent, Gaussian distributed, and equal in size for all fitted points. These assumptions are usually not justified (see, e.g., Refs. [52, 53] for the case of fitting power-law distributions), while at the same time the uncertainties are modified through the transformation of variables (such as using the log scale). Furthermore, quantifying the goodness-of-fit by using the correlation coefficient R^2 in these scales is insufficient to evaluate the validity of a given law. A high quality of the fit indicates a high correlation between data and model, but is unable to assign a probability for observations and thus it is not suited for a rigorous test of the law.

4.2 Likelihood Methods

A central quantity in the statistical analysis of data is the likelihood $\mathscr{L}(\mathbf{x}; \boldsymbol{\alpha})$ that the data $\mathbf{x}$ was generated by the model (with a set of parameters $\boldsymbol{\alpha}$).

(1) Fitting
When fitting a model (law) to data, the approach is to tacitly assume its validity and then search for the best parameters to account for the data. It corresponds to a search in the (multidimensional) parameter space $\boldsymbol{\alpha}$ of the law for the value $\hat{\boldsymbol{\alpha}}$ that maximize $\mathscr{L}$.

In laws of the first kind—as listed in Sect. 2—the quantity to be estimated from data is a probability distribution $P(\mathbf{x}; \boldsymbol{\alpha})$. The probability of an observation x_j is thus given by $P(x_j; \boldsymbol{\alpha})$. Assuming that all J observations are independent, the best parameter estimates $\hat{\boldsymbol{\alpha}}$ are the values of $\boldsymbol{\alpha}$ that maximize the log-likelihood

$$\log_e \mathscr{L} = \log P(x_1, x_2, \ldots, x_J; \boldsymbol{\alpha}) = \sum_{j=1}^{J} \log P(x_j; \boldsymbol{\alpha}), \qquad (6)$$

The need for Maximum Likelihood (ML) methods when fitting power-law distributions (such as Zipf's law) has been emphasized in many recent publications. We refer to the review article Ref. [54] and references therein for more details, and to Ref. [55] for fitting truncated distributions (e.g., due to cutoffs).

In laws of the second and third kind—as listed in Sect. 2—the quantity to be described y is a function $y = y_g(\mathbf{x}; \boldsymbol{\alpha})$. Fitting requires assumptions regarding the possible fluctuations in $y(\mathbf{x})$. One possibility is to assume Gaussian fluctuations with a standard deviation $\sigma(\mathbf{x})$. In this case, assuming again that the observations $\mathbf{x}$ are independent [56]

$$\log_e \mathscr{L} \sim - \sum_j \left(\frac{y(\mathbf{x}_j) - y_g(\mathbf{x}_j)}{\sigma(\mathbf{x}_j)} \right)^2, \qquad (7)$$

where the sum is over all observations j. The best estimated parameters $\hat{\boldsymbol{\alpha}}$ are obtained minimizing $\chi^2 = \sum_j (\frac{y(\mathbf{x}_j) - y_g(\mathbf{x}_j)}{\sigma(\mathbf{x}_j)})^2$, which maximizes (7). Least squares fitting is equivalent to Maximum Likelihood fitting only in the case of constant σ (independent of $\mathbf{x}$) [56].

(2) Model Comparison
The comparison between two different functional forms of the law ($m1$ and $m2$) is done comparing their likelihoods, e.g., through the log-likelihood ratio $\log_e \mathscr{L}_{m1}/\mathscr{L}_{m2}$ [57]. A value $\log_e \mathscr{L}_{m1}/\mathscr{L}_{m2} = 1$ (-1) means it is $e^1 = 2.718\ldots$ times more (less) likely that the data was generated by function $m1$ than by function $m2$ (see Ref. [54] for discussions on the significance of the log-likelihood ratio). If the two models have a different number of parameters, one can penalize the model with higher number of parameters, e.g., using the Akaike information criterion [58], calculating the Bayes factor by averaging (in the space of parameters) over the full posterior distribution [59], or using the principle of minimum description length [60].

(3) Validity
The probabilistic nature of linguistic laws requires statistical tests. One possible approach is to assume a null model (compatible with the linguistic law) and compare the fluctuations of finite-size outcomes of this model with the ones observed in the data. The probability (p-value) that the model generates fluctuations at least as extreme as the ones observed in the data may be used as a test of the validity of the model. A low p-value is a strong indication that the null model is violated and may be used to refute the law (e.g., if p-value < 0.01). Defining a measure of distance D between the data and the model, the p-value can be computed as the fraction of finite-size realizations of the model (assuming it is true) that show a distance $D' > D$. In

the case of probability distributions—linguistic laws of the first kind in Sect. 2—the distance D is usually taken to be the Kolmogorov–Smirnov distance (the largest distance between the empirical and fitted cumulative distributions). In the case of simple functions—linguistic laws of the second and third kind in Sect. 2—one can consider $D = \chi^2$.

Application: Menzerath–Altmann law

We applied the likelihood analysis summarized above to the case of the Menzerath–Altmann law introduced in Sect. 2. Our critical assumption here is that the law is intended to describe the average number of phonemes per syllable, y, computed over many words w with the same number of syllables x. Assuming the words are independent of each other, the uncertainty in $y(x)$ is thus the standard error of the mean given by $\sigma_y(x) = \sigma_w(x)/\sqrt{N(x)}$, where $\sigma_w(x)$ is the (empirical) standard deviation over the words with x-syllables and $N(x)$ is the number of such words.

In Fig. 2 and Table 2 we report the fitting, model comparison, and validity analysis for the Menzerath–Altmann law—Eq. (4)—and three alternative functions with the same number of parameters. The results show that two of the three alternative functions (shifted power law and stretched exponential) provide a better description than the proposed law, which we can safely consider to be incompatible with the data (p-value $< 10^{-5}$). Considering the two databases, the stretched exponential distribution provides the best description and is not refuted. These results depend strongly on the procedure used to identify phonemes and syllables (see Appendix).

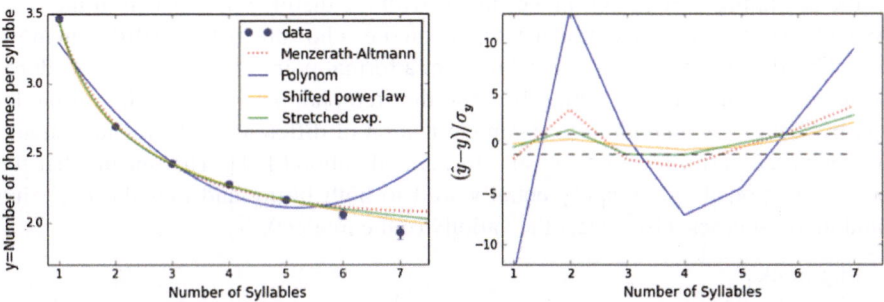

Fig. 2 Model comparison for the Menzerath–Altmann law. Data points are the average over all word (types) in a book (Moby Dick by H. Melville, as in Fig. 1). The curves show the best fits of the four alternative curves, as reported in Table 2. *Left plot* the data in the original scales, as in Fig. 1. *Right plot* the distance between the curves and the points $(\hat{y} - y)/\sigma_y$, where the uncertainty σ_y is the standard error of the mean

Table 2 Likelihood analysis of the Menzerath–Altmann law and three alternative functions

	Menzerath–Altmann (MA) $\alpha x^{\beta} \exp(-\gamma x)$	Shifted power law $\alpha(x+\beta)^{\gamma}$	Stretched exp. $\alpha \exp(\beta x)^{\gamma}$	Polynom $\alpha + \beta x + \gamma x^2$
Results for one book (Moby Dick by H. Melville)				
$(\hat{\alpha}, \hat{\beta}, \hat{\gamma})$	(3.3, −0.12, −0.051)	(2.8, −0.65, −0.19)	(1.5, 1.4, −0.51)	(3.9, −0.69, 0.066)
$\log_e \mathscr{L}_m/\mathscr{L}_{MA}$	0	**33**	25	−475
p-value	$<10^{-5}$	**0.611**	**0.064**	$<10^{-5}$
Results for English Wikipedia				
$(\hat{\alpha}, \hat{\beta}, \hat{\gamma})$	(3.2, −0.45, −0.064)	(2.8, −0.70, −0.18)	(1.6, 1.5, −0.60)	(3.8, −0.64, 0.061)
$\log_e \mathscr{L}_m/\mathscr{L}_{MA}$	0	11	**49**	−1898
p-value	$<10^{-5}$	2×10^{-5}	**0.93**	$<10^{-5}$

The parameters $(\hat{\alpha}, \hat{\beta}, \hat{\gamma})$ that maximize the likelihood $\mathscr{L}_m$ of model m were computed using the downhill simplex algorithm (using the Python library scipy). The reported p-value corresponds to the fraction of random realizations with a χ^2 larger than the observed χ^2. In each realization, one point $y^{\dagger}(x)$ was generated at each x from a Gaussian distribution centered at the model prediction $y_m(x)$ with a standard deviation $\sigma_y(x)$ given by the data. The best models and the results with $p > 0.01$ are shown in bold face

4.3   Critical Discussion

In the next paragraphs we critically discuss the likelihood approach considering the example of Zipf's law.

Fitting as Model Comparison

In the beginning of this section we started with the distinction between fitting (i.e., fixing free parameters) and model comparison (i.e., choosing between different models). This division is didactic [54], but from a formal point of view both procedures correspond to hypothesis testing because the free parameters of one fitting model can be thought as a *continuous* parameterization of different models which should be compared and selected according to their likelihood [61]. This means that the points mentioned below apply equally well to both fitting and hypothesis testing (and, in most cases, also to test the validity of the models).

Fitting Ranks

Power-law fitting recipes [54]—employed for linguistic [46] and nonlinguistic problems—suggest to fit Zipf's law using the distribution of frequencies $P(f)$ given in Eq. (3). However, it is also possible to use the rank formulation (2) [22] because the frequency of ranks $f(r)$ is normalized $\sum_r f(r) = 1$ and can thus be interpreted as a probability distribution. However, a drawback in fitting $f(r)$ is that the process of ranking introduces a bias in the estimator [62, 63]. For instance, consider a finite sample from a true Zipf distribution containing ranks $r = 1, \ldots, \infty$. Because of statistical fluctuations, some of the rankings will be inverted (or absent) so that when we rank the words according to the observations obtain ranks different from the ones drawn. This effect introduces bias in our estimation of the parameters (overestimating the quality of the fit). However, the words affected by this bias are the ones with largest ranks, which contribute very little to the estimation of the parameters of Zipf's law (as discussed below). Therefore, we expect this bias to become negligible even for moderately large sample sizes.

Representation Matters

Equivalent formulations of the linguistic laws lead to *different* statistical analysis and conclusions [62, 63]. One example of this point is the use of transformations before the fitting is performed, such as the linear fit of Zipf's law in logarithmic scale discussed in Sect. 4.1. The variables used to represent the linguistic law are also crucial when likelihood methods are used, as discussed above for the case of Zipf's law represented in $f(r)$ or $P(f)$. While asymptotically ($N \to \infty$) these formulations are equivalent, the likelihood computed in both cases is different. In the likelihood of $P(f)$, an observation corresponds to the frequency of a word *type*. This means that the most frequent words in the database count the same as words appearing only once (the hapax legomena). In practice, the part of the distribution that matters the most in the fitting (and in the likelihood) are the words with very few counts, which contribute very little to the total text. In the likelihood of $f(r)$ the observational quantity is the rank r of each occurrence of the word meaning that each word *token* counts the same. This means that the frequent words contribute more and the fitting of $f(r)$ is robust against rare words. Linear regression in log–log plot counts every point in the plot the same and, since there are more points for large r, low-frequency words dominate the fit. Using logarithmic binning, as suggested in Ref. [52], equalize the importance of words across $log(r)$. In summary, while fitting a straight line in log–log scale using logarithmic binning gives the same value for words across the full spectrum (in a logarithmic scale), the statistical rigorous methods of Maximum Likelihood will be dominated either by the most frequent (in case of fitting in $f(r)$) or least frequent (in case of fitting in $P(f)$) words.

Beyond Zipf's law, the reasoning above shows that even if asymptotically (i.e., infinite data) different formulations of a law are equivalent, the representation in which we test the law matters because it assumes a sampling process of the data. This in turn leads to different results when applied to finite and often noisy data and has to be taken into account when interpreting the results.

Application: Fitting Zipf's law

In Fig. 3 and Table 3 we compare the different fitting methods described above. The visual agreement between data and the fitted curves reflects the different weights given by the methods to different regions of the distribution as discussed above (high-frequency words for $f(r)$ and low-frequency words for the other two cases). Not surprisingly, Table 3 shows that the estimated exponent α varies from method to method. This variation is larger than the variation across different databases. Large values of R^2 computed in the linear fit, usually interpreted as an indication of good fitting, are observed also when the p-values are very low.

Correlated Samples

The failure of passing significance tests for increasing data size is not surprising because any small deviation from the null model becomes statistically significant.

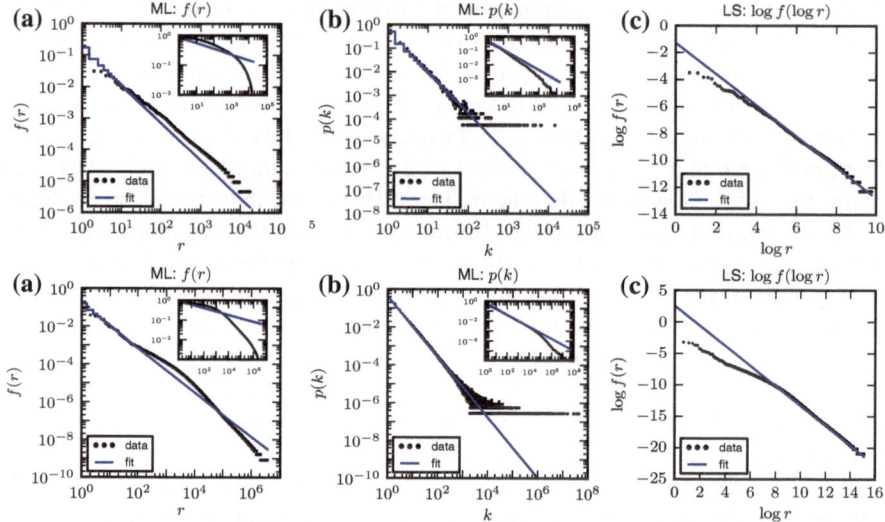

Fig. 3 Comparison of the Zipf's law obtained using three different fitting methods. Results are shown for one book (Moby Dick by H. Melville, *top row*) and for the complete English Wikipedia (*bottom row*). Data is fitted using Maximum Likelihood (ML) in the frequency rank $f(r)$ (*left*), ML in the frequency distribution $P(f) \sim p(k)$ (*center*), and least square (LS) in the log f versus log r representation (*right*). *Insets* show the cumulative distributions. See Table 3 for the parameter $\hat{\alpha}$ and significance test of the fits. In the plot in the *center*, instead of $P(f)$ we use the distribution the unnormalized frequency $p(k)$ (i.e., k is the number of occurrences of a word in the database). For ML fits, we used a discrete power law in $f(r)$ and $p(k)$ with support in $[1, \infty)$ (exponents were obtained using the downhill simplex algorithm of the Python library scipy). For the LS fit, we used a continuous *straight line* in log f (log r) for all $r > 0$ [56]

Table 3 Zipf's law exponent obtained using different fitting methods, see Fig. 3

Book	Rank: $f(r)$		Frequency: $P(f)$		Linear: $\log f(\log r)$	
	$\hat{\alpha}_Z$	p-value	$\hat{\alpha}_Z$	p-value	$\hat{\alpha}_Z$	R^2
Alice's Adventures in Wonderland (L. Carroll)	1.22	$<10^{-4}$	1.46	$<10^{-4}$	1.21	0.97
The Voyage Of The Beagle (C. Darwin)	1.20	$<10^{-4}$	1.59	$<10^{-4}$	1.29	0.97
The Jungle (U. Sinclair)	1.21	$<10^{-4}$	1.45	$<10^{-4}$	1.22	0.98
Life On The Mississippi (M. Twain)	1.20	$<10^{-4}$	1.38	$<10^{-4}$	1.16	0.98
Moby Dick; or The Whale (H. Melville)	1.19	$<10^{-4}$	1.38	$<10^{-4}$	1.15	0.98
Pride and Prejudice (J. Austen)	1.21	$<10^{-4}$	1.66	$<10^{-4}$	1.35	0.98
Don Quixote (M. Cervantes)	1.21	$<10^{-4}$	1.70	$<10^{-4}$	1.38	0.98
The Adventures of Tom Sawyer (M. Twain)	1.21	$<10^{-4}$	1.29	$<10^{-4}$	1.12	0.98
Ulysses (J. Joyce)	1.18	$<10^{-4}$	1.15	$<10^{-4}$	1.03	0.97
War and Peace (L. Tolstoy)	1.20	$<10^{-4}$	1.84	$<10^{-4}$	1.44	0.97
English Wikipedia	1.17	$<10^{-4}$	1.60	$<10^{-4}$	1.58	0.99

In the fit of $P(f)$ (frequency) we obtain $\hat{\alpha}_Z^\dagger$ and calculate $\hat{\alpha}_Z = 1/(\hat{\alpha}_Z^\dagger - 1)$, see Eqs. (2) and (3). English version of the books were obtained from the Project Gutenberg, see Appendix

A possible conclusion emerging from these analysis is that power-law distributions are not as widely valid as previously claimed (see also Refs. [54, 64]), but often are better than alternative (simple) descriptions (see our previous publication Ref. [22] in which we consider two-parameter generalizations of Zipf's law). The main criticism we have on this widely used framework of analysis is that it ignores the presence of correlations in the data: the computation of the likelihood in Eq. (6) assumes independent observations. Furthermore, this assumption leads to an underestimation of the expected fluctuations (e.g., KS distance) in the calculation of the p-value when assessing the validity of the law. It is thus unclear in which extent a negative result in the validity test (e.g., p-value $\ll 0.01$) is due to a failure of the proposed law or, instead, is due to the violation of the hypothesis of *independent* sampling. This hypothesis is known to be violated in texts [7, 27]: the sequence of words and letters are obviously related to each other. In Fig. 4 we show that these correlations affect the estimation of the frequency f of individual words. Fluctuations are much larger than expected not only from the independent random usage of words (Poisson or bag-of-words models) but also from a null model in which burstiness is included [25, 27]. Altogether, this shows that the independence assumption—used to write the likelihood (6)—is strongly violated. For the analysis of Zipf's law discussed above, the obvious correlation throughout (the tokens of) a book directly affects the (rank) analysis based on $f(r)$. Also the (frequency) analysis based on $P(f)$ is strongly affected by correlations because the estimation of the frequency f of a word is always performed in a finite sample (finite book) which is analogous to an $n < N$ in Fig. 4.

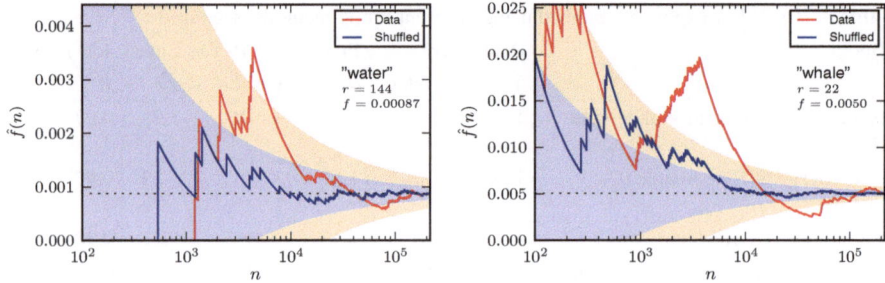

Fig. 4 Estimation of the frequency of a word in the first n word tokens of a book (Moby Dick by H. Melville). The *red curve* corresponds to the actual observation (word "water" in the *left* and word "whale" in the *right*) and the *blue curve* to the curve measured in a version of the book in which all word tokens were randomly shuffled. The shaded regions show the expected fluctuations ($\pm 2\sigma$) assuming that the probability of using the word is given by the frequency of the word in the whole book ($f(n = N)$ and that: (i) usage is random (*blue region*)—see also Ref. [7] or (ii) the time between successive usages of the word is drawn randomly from a stretched exponential distribution with exponent $\beta = 0.5$, as proposed in Ref. [25]

One approach to take into account correlations is to estimate a time for which two observations are independent, and then consider observations only after this time (a smaller effective sample size). Alternative approaches considered statistical tests for specific classes of stochastic processes (correlated in time) [65] or based on estimations of the correlation coming from the data [66]. The application of these methods to linguistic laws is not straightforward because these methods fail in cases in which no characteristic correlation time exist. Books show such long-range correlations [29], also in the position of individual words in books [28, 30], in agreement with the observations reported in Fig. 4. More generally, correlations lead to a slower convergence to asymptotic values and it is thus possible to create processes of text generation that comply to a linguistic law asymptotically but that (in finite samples) violate statistical tests based on independent sampling. The problem affects also model comparison and fitting because these problems are also based on the likelihood (in these cases, correlation affects all models and therefore it is unclear the extent in which it impacts the choice of the best model).

5   Relation Between Laws

In view of the different laws proposed to describe text properties, a natural question is the relationship between them (e.g., whether one law can be derived from another or whether there are generative processes that account for more than one law simultaneously). For instance, Ref. [30] clarifies how the long-range correlation of texts is related to the skewed distribution of recurrence time between words [2, 25–27] (a consequence of burstiness [8, 67]). Another well-known relation is the connection between Heaps' law and Zipf's law [20, 22–24, 43] (see Refs. [24, 31, 32, 67] for other examples). Here again the importance of fluctuations and an underlying null model is often neglected.

The need for a null model is evident if we consider a text in which all possible words appear once in the very beginning of the text, violating Heaps' law, even though their frequency over the full text is still compatible with Zipf's law. A typical null model is to consider that every word is used independently from the others with a probability equal to its global frequency. This probability is usually taken to be constant throughout the text (Poisson process), but alternative formulations considering time-dependent frequencies lead to similar results. For this generative model, Zipf's law (2) leads to a Heaps' law (5) with parameters $\alpha_H = 1/\alpha_Z$ [22]. Similar null models are implicitly or explicitly assumed in different derivations [20, 22–24, 43].

Figure 5 shows that the connection between Zipf's and Heaps' law using the independent usage of words fails to reproduce the fluctuations observed in data. In particular, the fluctuations around the average vocabulary size V (Heaps' law) scales linearly with N, and not as $\sqrt{N}$ as predicted by the independence assumption (through the central limit theorem). In Ref. [24] we have shown that this scaling—also known as Taylor's law [68]—is a result of correlations in the usage of different words induced by the existence of topical structures inside and across books.

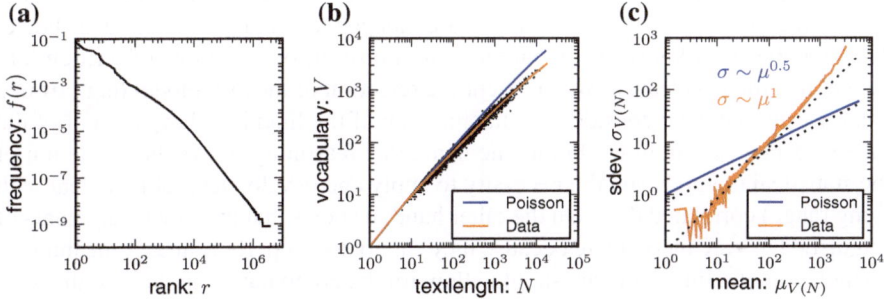

Fig. 5 Relation between Zipf's law and Heaps' law in the English Wikipedia. Fixing the rank-frequency distribution of the complete English Wikipedia—shown in panel (**a**)— and assuming each word to follow a Poisson process (i.e., to be used randomly) with fixed frequency $f(r)$, one obtains the *blue curve* for the Heaps' law in (**b**). Considering each Wikipedia article separately—as shown by *black dots* in (**b**)—we estimate in a moving window centered in N the average $\mu_{V(N)}$ and standard deviation $\sigma_{V(N)}$ over all articles in the window. The dependence of $\mu_V(N)$ on N is shown in (**b**) by a *solid line*. The dependence of $\sigma_V(N)$ on $\mu_V(N)$ is shown in (**c**) and reveals a different scaling than the one predicted by the Poisson model. Figure adapted from Ref. [24]

6 Discussion

It is common to find claims that a particular linguistic law is valid in a language or corpus. A closer inspection for the statistical support of these claims is often disappointing. In this chapter we performed a critical discussion of linguistic laws, the sense in which they can be considered valid, and the extent in which the evidence support its validity. We argued that linguistic laws have to be interpreted in a statistical sense. Therefore, model selection (also fitting) and the compatibility to data have to be performed computing statistical tests based on the likelihood (plausibility) of the observations. Scientists still have to choose the statistical test and the additional assumptions (not contained in the original law) which are more appropriate to address each specific question. The analysis presented above is intended to show that these choices matter and should be carefully discussed. The picture that emerges from the straight applications of the statistical tests above is that: (i) the linguistic laws are often the best simple description of the data, but (ii) the data is not generated according to it so that in a strict sense the validity of the law is falsified. This interpretation suggests that linguistic laws are useful and capture some of the ingredients seen in language, but are unable to describe the observations in full detail even in the limit of large texts (possibly because of the existence of additional processes ignored by the law).

The main limitation of the methods we described, and thus of the conclusions summarized above, is that they were based not only on the statement of the law but also on the hypothesis that observations are independent and identically distributed. This hypothesis is known to be violated in almost all observations of written language. It is thus unclear in which extent the rejection of the null model (e.g., from a small p-value) can be considered a falsification of the linguistic law, and not of the independency assumption. On the one hand, this reasoning shows the limitation of the statistical methods and the necessity to apply and develop tests able to deal with (long range) correlated data. On the other hand, it shows that the usual statements of linguistic laws are incomplete because they cannot be properly tested. A meaningful formulation of a linguistic law should allow for the computation of the likelihood of the observations, e.g., it should be accompanied by a prediction of the fluctuations, a generative model for the relevant variables, or, ultimately, a model for the generation of texts. Such models are usually interpreted as an explanation of the origin of the laws [10, 11, 19] and are absent from the statement of the linguistic laws, despite the fact that Herdan already drew attention to this point [1]: *The quantities which we call statistical laws being only expectations, they are subject to random fluctuations whose extent must be regarded as part of the statistical law.* In the same sense that a scientific law cannot be judged separated from a theory, linguistic laws are only fully defined once a generative process is given. The existence of long-range correlations, burstiness, and topical variations lead to strong fluctuations in the estimations of observables in texts, including the quantities described by linguistic laws.

Our findings have consequences to applications in information retrieval and text generation. For instance, our results show that strong fluctuations around specific laws are observed and that results obtained using the independence assumption (e.g., bag-of-words models) have a limited applicability. Therefore, statistical laws should not be imposed too strictly in the generation of artificial texts or in the analysis of unknown databases. Large fluctuations are as much a characteristic of language as the laws themselves and therefore the creativity in the generation of texts is much larger than the one obtained if laws are imposed as strict constraints.

Finally we would like to mention that our conclusions apply also to statistical laws beyond linguistics. Invariably, the increase of data size leads to a rejection of null models, e.g., many recent works emphasize that claims of power-law distributions do not survive rigorous statistical tests [17, 54, 64]. However, the statistical tests employed in these references, and in most likelihood-based analysis, rely on the independence assumption of the observations (known to be violated in many of the treated cases). Nevertheless, we are not aware that this point has been critically discussed in the large number of publications on power-law fitting. The crucial role of mechanistic models in the fitting and statistical analysis of scaling laws was emphasized in Ref. [69] for urban economic data.

Acknowledgments We thank A. Corral, A. Deluca, R. Ferrer-i-Cancho F. Font-Clos, and R. Guimerá for insightful discussions.

Appendix

The books listed in Table 3 were obtained from Project Gutenberg (http://www.gutenberg.org). The books and data filtering are the same as the ones used in Ref. [30] (see the Supplementary information of that paper for further details). We removed capitalization and all symbols except the letters "a–z", the number "0–9", the apostrophe, and the blank space. A string of symbols between two consecutive blank spaces was considered to be a word.

The English Wikipedia data was obtained from Wikimedia dumps (http://dumps.wikimedia.org/). The filtering was the same as the one used in Ref. [24], in which we removed capitalization and kept only those words (i.e., sequences of symbols separated by blank space) which consisted exclusively of the letters "a–z" and the apostrophe.

The computation of Menzerath–Altmann law appearing in Figs. 1, 2, and Table 2 was done starting from the unique words (word type) in the database discussed in the previous paragraphs. For each word w we applied the following steps:

1. Lemmatize using the WordNetLemmatizer (http://wordnet.princeton.edu in the NLTK Python package http://www.nltk.org/).
2. Count the number of syllables x_w based on the *Moby Hyphenation List by Grady Ward*, available at http://www.gutenberg.org/ebooks/3204.
3. Count the number of phonemes z_w based on *The CMU Pronouncing Dictionary*, version 0.7b available at www.speech.cs.cmu.edu/cgi-bin/cmudict.

For the book *Moby Dick* by H. Melville, this procedure allowed to compute x_w and z_w for 11, 595 words, 66 % of the total number of words (before lemmatization). For the Wikipedia, we obtain 60, 749 words, 1.7 % of the total number. The low success in Wikipedia is due to the size of the database (large number of rare words) and the results depend more strongly on the procedure described above than on the database itself.

References

1. Herdan, G.: Quantitative Linguistics. Butterworth Press, Oxford (1964)
2. Zipf, G.K.: The Psycho-Biology of Language. Routledge, London (1936). Id., Human Behavior and the Principle of Least Effort. Addison-Wesley Press, Oxford (1949)
3. Köhler, R., Altmann, G., Piotrowski, R.G. (eds.): Quantitative Linguistik. Ein internationales Handbuch. Quantitative Linguistics. An international Handbook. (=HSK27). de Gruyter, Berlin (2005)
4. Köhler, R., Altmann, G., Grzybek, P. (eds.): Quantitative Linguistics, De Gruyer Mouton. www.degruyter.com/view/serial/35295. Accessed 6 Feb 2015
5. Glottopedia: the free encyclopedia of linguistics. http://www.glottopedia.org/index.php/Laws. Accessed 17 Dec 2014
6. Enciclopedia entry: laws in quantitative linguistics. http://lql.uni-trier.de. Accessed 3 Dec 2014
7. Harald Baayen, R.: Word Frequency Distributions. Kluwer Academic Publishers, Dordrecht (2001)
8. Zanette, D.H.: Statistical patterns in written language (2014). arXiv:1412.3336
9. Barbieri, G., Pachet, F., Roy, P., Degli Esposti, M.: Markov constraints for generating lyrics with style. In: 20th European Conference on Artificial Inteligence – ECAI, IOS Press, Amsterdam (2012)
10. Mitzenmacher, M.: A brief history of generative models for power law and lognormal distributions. Internet Math. **1**, 226 (2004)
11. Newman, M.E.J.: Power laws, Pareto distributions and Zipfs law. Contemp. Phys. **46**, 323 (2005)
12. Mandelbrot, B.: On the theory of word frequencies and on related Markovian models of discourse. In: Structure of Language and Its Mathematical Aspects: Proceedings of Symposia in Applied Mathematics, vol. XII. American Mathematical Society, Providence (1961)
13. Altmann, G.: Prolegomena to Menzerath's law. Glottometrika **2**, 1 (1980)
14. Cramer, I.: The parameters of the Altmann-Menzerath law. J. Quant. Linguist. **12**, 41 (2005)
15. Egghe, L.: Untangling Herdan's law and Heaps' law : mathematical and informetric arguments. J. Am. Soc. Inf. Sci. Technol. **58**, 702 (2007)
16. Simon, H.A.: On a class of skew distribution functions. Biometrika **42**, 425 (1955)
17. Li, W.: Zipf's law everywhere. Glottometrics **5**, 14 (2002)
18. Zanette, D., Montemurro, M.: Dynamics of text generation with realistic Zipf's distribution. J. Quant. Linguist. **12**, 29 (2005)
19. Piantadosi, S.T.: Zipfs word frequency law in natural language: a critical review and future directions. Psychon. Bull. Rev. **21**, 1112 (2014)
20. Lü, L., Zhang, Z.-K., Zhou, T.: Zipf's law leads to Heaps' law: analyzing their relation in finite-size systems. PLOS One **5**, e14139 (2010)
21. Petersen, A.M., Tenenbaum, J.N., Havlin, S., Stanley, H.E., Perc, M.: Languages cool as they expand: allometric scaling and the decreasing need for new words. Sci. Rep. **2**, 943 (2012)
22. Gerlach, M., Altmann, E.G.: Stochastic model for the vocabulary growth in natural languages. Phys. Rev. X **3**, 021006 (2013)

23. Font-Clos, F., Boleda, G., Corral, A.: A scaling law beyond Zipf's law and its relation to Heaps' law. New J. Phys. **15**(9), 093033 (2013)
24. Gerlach, M., Altmann, E.G.: Scaling laws and fluctuations in the statistics of word frequencies. New J. Phys. **16**, 113010 (2014)
25. Altmann, E.G., Pierrehumbert, J.B., Motter, A.E.: Beyond word frequency: bursts, lulls, and scaling in the temporal distributions of words. Plos One **4**, e7678 (2009)
26. Corral, A., Ferrer-i-Cancho, R., Boleda, G., Diaz-Guilera, A.: Univeral complex structures in written language. arXiv:0901.2924
27. Lijffijt, J., Papapetrou, P., Puolamäki, K., Mannila, H.: Analyzing word frequencies in large text corpora using inter-arrival times and bootstrapping. Machine Learning and Knowledge Discovery in Databases. Lecture Notes in Computer Science, vol. 6912, p. 341. Springer, Berlin (2011)
28. Damerau, F.J., Mandelbrot, B.: Tests of the degree of word clustering in samples of written English. Linguistics **102**, 58–72 (1973)
29. Schenkel, A., Zhang, J., Zhang, Y.: Long range correlation in human writings. Fractals **1**, 47 (1993)
30. Altmann, E.G., Cristadoro, G., Degli Esposti, M.: On the origin of long-range correlations in texts. PNAS **109**, 11582 (2012)
31. Ebeling, W., Pöschel, T.: Entropy and long-range correlations in literary English. Europhys. Lett. **26**, 24 (1994)
32. Debowski, L.: On Hilberg's law and its links with Guiraud's law. J. Quant. Linguist. **13**, 81–109 (2006)
33. Piantadosi, S.T., Tily, H., Gibson, E.: Word lengths are optimized for efficient communication. PNAS **108**, 3526 (2011)
34. Solé, R.V., Corominas-Murtra, B., Valverde, S., Steels, L.: Language networks: their structure, function and evolution. Complexity **15**, 20 (2009)
35. Choudhury, M., Mukherjee, A.: The structure and dynamics of linguistic networks. Dynamics On and Of Complex Networks, pp. 145–166. Springer, Boston (2009)
36. Baronchelli, A., Ferrer-i-Cancho, R., Pastor-Satorras, R., Chater, N., Christiansen, M.H.: Networks in cognitive science. Trends Cogn. Sci. **17**, 348 (2013)
37. Cong, J., Liu, H.: Approaching human language with complex networks. Phys. Life Rev. **11**, 598 (2014)
38. Constrained writing, in Wikipedia. http://en.wikipedia.org/wiki/Constrained_writing. Accessed 3 Dec 2014
39. Benford, F.: The law of anomalous numbers. Proc. Am. Philos. Soc. **78**, 551 (1938)
40. Main, I.G., Li, L., McCloskey, J., Naylor, M.: Effect of the Sumatran mega-earthquake on the global magnitude cut-off and event rate. Nat. Geosci. **1**, 142 (2008)
41. Amancio, D.R., Altmann, E.G., Rybski, D., Oliveira Jr., O.N., Costa, L.D.F.: Probing the statistical properties of unknown texts: application to the Voynich manuscript. PLOS One **8**, e67310 (2013)
42. Febres, G., Jaffé, K., Gershenson, C.: Complexity measurement of natural and artificial languages. Complexity (2014). doi:10.1002/cplx.21529
43. Bernhardsson, S., da Rocha, L.E.C., Minnhagen, P.: Size-dependent word frequencies and translational invariance of books. Phys. A **389**, 330 (2010)
44. Williams, J.R., Bagrow, J.P. Danforth, C.M., Dodds, P.S.: Text mixing shapes the anatomy of rank-frequency distributions: a modern Zipfian mechanics for natural language (2014). arXiv:1409.3870
45. Baixeries, J., Elevag, B., Ferrer-i-Cancho, R.: The evolution of the exponent of Zipf's law in language ontogeny. PLOS One **8**, e53227 (2013)
46. Jäger, G.: Power laws and other heavy-tailed distribution in linguistic typology. Adv. Compl. Syst. **15**, 1150019 (2012)
47. Ferrer-i-Cancho, R., Elevag, B.: Random texts do not exhibit the real Zipf's law-like rank distribution. PLOS One **5**, e9411 (2010)

48. Corominas-Murtra, B., Fortuny, J., Solé, R.V.: Emergence of Zipfs law in the evolution of communication. Phys. Rev. E **83**, 036115 (2011)
49. Ferrer-i-Cancho, R.: Optimization models of natural communication (2014). arXiv:1412.2486
50. Marsili, M., Mastromatteo, I., Roudi, Y.: On sampling and modeling complex systems. J. Stat. Mech. **2013**, P09003 (2013)
51. Peterson, J., Dixit, P.D., Dill, K.: A maximum entropy framework for nonexponential distributions. PNAS **110**, 20380 (2013)
52. Goldstein, M.L., Morris, S.A., Yen, G.G.: Problems with fitting to the power-law distribution. Eur. J. Phys. B **41**, 255–258 (2004)
53. Bauke, H.: Parameter estimation for power-law distributions by maximum likelihood methods. Eur. J. Phys. B **58**, 167–173 (2007)
54. Clauset, A., Shalizi, C.R., Newman, M.E.J.: Power-law distributions in empirical data. SIAM Rev. **51**, 661–703 (2009)
55. Deluca, A., Corral, A.: Fitting and goodness-of-fit test of non-truncated and truncated power-law distributions. Acta Geophys. **61**, 1351–1394 (2013)
56. Hastie, T., Tibshirani, R., Friedman, J.: The Elements of Statistical Learning. Springer, New York (2009)
57. Burnham, K.P., Anderson, D.R.: Model Selection and Multimodal Inference: A Practical Information-Theoretic Approach. Spinger, New York (2002)
58. Akaike, H.: A new look at the statistical model identification. IEEE Trans. Autom. Control **19**, 716–723 (1974)
59. Kass, R.E., Raftery, A.E.: Bayes factors. J. Am. Stat. Assoc. **90**, 773–795 (1995)
60. Grünwald, P.D.: Minimum Description Length Principle. MIT Press, Cambridge (2007)
61. Jaynes, E.T.: Probability Theory: The Logic of Science. Oxford University Press, Oxford (2003)
62. Günther, R., Levitin, L., Schapiro, B., Wagner, P.: Zipf 's law and the effect of ranking on probability distributions. Int. J. Theor. Phys. **35**, 395 (1996)
63. Cristelli, M., Batty, M., Pietronero, L.: There is more than a power law in Zipf. Sci. Rep. **2**, 812 (2012)
64. Stumpf, M.P.H., Porter, M.A.: Critical truths about power laws. Science **335**, 665–666 (2012)
65. Weiss, M.S.: Modification of the Kolmogorov-Smirnov statistic for use with correlated data. J. Am. Stat. Assoc. **73**, 872–875 (1978)
66. Chicheportiche, R., Bouchaud, J.-P.: Goodness-of-fit tests with dependent observations. J. Stat. Mech.: Theory Exp. **2011**, P09003 (2011)
67. Serrano, M.A., Flammini, A., Menczer, F.: Modeling statistical properties of written text. PlOS One **4**, e5372 (2009)
68. Eisler, Z., Bartos, I., Kertész, J.: Fluctuation scaling in complex systems: Taylor's law and beyond. Adv. Phys. **57**, 89–142 (2008)
69. Louf, R., Barthelemy, M.: Scaling: lost in the smog. Environ. Plan. B: Plan. Des. **41**, 767 (2014)

Complexity and Universality in the Long-Range Order of Words

Marcelo A. Montemurro and Damián H. Zanette

Abstract As is the case of many signals produced by complex systems, language presents a statistical structure that is balanced between order and disorder. Here we review and extend recent results from quantitative characterisations of the degree of order in linguistic sequences that give insights into two relevant aspects of language: the presence of statistical universals in word ordering, and the link between semantic information and the statistical linguistic structure. We first analyse a measure of relative entropy that assesses how much the ordering of words contributes to the overall statistical structure of language. This measure presents an almost constant value close to 3.5 bits/word across several linguistic families. Then, we show that a direct application of information theory leads to an entropy measure that can quantify semantic structures and extract keywords from linguistic samples, even without prior knowledge of the underlying language.

1 Introduction

There are only a few known cases of systems that have naturally evolved to encode complex information. For much of Earth's history the chemical language of the genetic code has been the prime example. And although there is evidence that non-human animal species have also developed means of non-trivial communication [1–3], it was not until the emergence of human language that the last major transition in evolution took place [4]. This human faculty evolved as an efficient system capable of transmitting sophisticated messages between different brains, becoming closely linked to our higher mental functions [5].

As a carrier of highly complex information, human language must operate under the competing requirements of allowing high information rate and at the same time

M.A. Montemurro (✉)
Faculty of Life Sciences, The University of Manchester, Manchester, UK
e-mail: m.montemurro@manchester.ac.uk

D.H. Zanette
Centro Atómico Bariloche e Insituto Balseiro, San Carlos de Bariloche,
Río Negro, Argentina
e-mail: zanette@cab.cnea.gov.ar

© Springer International Publishing Switzerland 2016   27
M. Degli Esposti et al. (eds.), *Creativity and Universality in Language*,
Lecture Notes in Morphogenesis, DOI 10.1007/978-3-319-24403-7_3

being robust under communication errors. These constraints, of both novelty and redundancy, contribute to shape a statistical structure in linguistic sequences that pose them at a balanced point between order and disorder. Recent advances on the analysis of language with methods and concepts from statistical physics and information theory have disclosed a rich structure at various level of linguistic organisation [6]. Here we review and extend recent results on the characterisation of linguistic order by means of novel entropy measures.

In the first part we discuss a measure of relative entropy that specifically quantifies the degree of order in word patterns. We show that this measure presents a universal value when it is evaluated on language samples belonging to 24 linguistic families. While in their evolutionary history different languages have developed a diverse range of underlying rules and vocabularies, the data suggest that their evolution and diversification were constrained to have an almost constant measure of relative entropy.

In the second part, we use another entropy measure that is capable of quantifying patterns in word distribution that are closely linked to the semantic role of words. Without any prior linguistic knowledge about the underlying language, we show that it is possible to extract the words that are most closely related to the semantic content of a text and, moreover, disclose semantic relationships between them.

2 Universality in the Entropy of Word Ordering

We may ask the question whether the precise balance between structure and randomness in linguistic sequences depends on features of specific languages, or instead represents some universal aspect of the human language faculty. Some linguists have put forward the hypothesis that even all languages share some basic structural features indicative of cognitive constraints [7–10], while others have challenged the existence of such linguistic universals [11] or argued that cultural, rather than cognitive traits, are responsible for widespread similarities across some linguistic families [12]. Recently, it has been shown that some patterns in word ordering, like the basic arrangement of subject, verb, and object in sentences, depends on the evolutionary and phylogenetic history of language [13]. Despite the controversy on the presence of linguistic universals at the level of language structure, quantitative aspects of language presenting universal characteristics have been established. The two best-known examples are Zipf's [14] and Heap's [15] laws in language, which refer to universal features related to word frequencies [16–19]. However, quantitative assessment of universality of word ordering in language are rarer.

The entropy of a symbolic stochastic source is linked to the predictability of the subsequent outcomes of the sequence when past values are known. A high predictability of future values will entail a low level of surprise in the new symbols, and hence a low entropy. Conversely, a perfectly random sequence will have the highest possible surprise in its symbols, and thus will be characterised by high entropy. Although language sequences are not produced by a stochastic source, it is generally assumed that a large collection of language samples represent an ensemble with enough consistency in its statistical structure to allow the application of the standard formalism of information theory. However, one serious hurdle in computing the entropy of language based on the estimation of block probabilities is the presence of long-range correlations that span from hundreds to thousands of words [20–24]. The sample size that would be needed to estimate the required probabilities grows exponentially with block length, thus quickly rendering insufficient any available linguistic source. One way in which this problem can be overcome is through the link between entropy and predictability. Non-parametric estimations of the entropy of language based on guessing games—where subjects have to predict future characters based on the past history of the linguistic sequence—were shown to yield useful results even with moderate sample sizes [25, 26]. Along similar lines, the degree of predictability in a sequence determines how much it could be compressed by a lossless compression method. Highly predictable sequences can be compressed further than more random ones. More rigorously, it can be shown that under the assumptions of stationarity and ergodicity the entropy rate of a stochastic source is a lower bound to the length per symbol of any encoding of it [27]. This suggests an approach to estimate the entropy of a symbolic sequence based on the use of efficient lossless compression algorithms. Many of the practical applications of these ideas are based on the complexity measure [28] and compression algorithms [29, 30] proposed by A. Lempel and J. Ziv, which rely on the estimation of redundancy by matchings between future and past substrings in a symbolic sequence. More recently, methods that estimate the entropy directly by string matching without attempting to compress the symbolic sequence have also been shown to be efficient [31, 32]. Implementations of these methods have proved to work well for symbolic sequences even in the presence of long-range correlations as those found in language [33–36], and without requiring very long sequences in order to converge [37].

In [38] we carried out an analysis of 7,077 texts from 8 languages from 5 linguistic families and one language isolate[1] to assess the contribution to word ordering to the statistical structure of language. The entropy, H, was estimated for every text by means of methods derived form compression algorithms and string matching. In order to account for the contribution to linguistic structure that comes only from word frequencies and irrespective of word ordering we also computed the entropy of a randomly shuffled version of the texts, H_s. To calculate the entropy of the disordered

---

[1] All but the Old Egyptian and Sumerian text sources were obtained from the Project Gutenberg e-text repository (www.gutenberg.org). The Old Egyptian texts were obtained from the page maintained by Dr. Mark-Jan Nederhof at the University of St Andrews (www.cs.st-andrews.ac.uk/mjn/egyptian/texts/) as transliterations from the original hieroglyphs. The Sumerian texts were downloaded from The Electronic Text Corpus of Sumerian Literature (www-etcsl.orient.ox.ac.uk/) and consisted of transliterations of the logo-syllabic symbols.

texts, we first computed the total number of possible arrangements between the words in a given text, as follows:

$$\Omega = \frac{N!}{\prod_{j=1}^{K} n_j!},$$ (1)

where K is the size of the vocabulary and n_j represents the number of instances of the word with index $j = 1 \ldots K$. Then, the entropy can be estimated à la Boltzmann, as follows:

$$H_s = \frac{1}{N} \log_2 \Omega.$$ (2)

Since the random texts lack any linguistic structure beyond word frequencies, the entropy will be larger than that of the original sequence, H. Therefore, one way to quantify the impact of the ordering of words is by means of a relative entropy measure, defined as $D_s = H_s - H$. In [38] it is shown that for sufficiently long sequences this quantity is equivalent to the Kullback–Leibler (KL) divergence between the original and disordered sequences. This measure quantifies the degree of order in a linguistic sequence beyond that contributed by word frequencies alone.

Figure 1 shows an example of the results obtained for three corpora of languages that differ significantly in their structure: Chinese (Sino-Tibetan), English (Indo-European), and Finnish (Finno-Ugric). In each panel the rightmost distribution corresponds to the entropy of the random texts, H_s, which only accounts for the contribution of word frequencies. While for English and Chinese the value of the distribution of H_s peaks around 9 bits/word, for Finnish it is close to 11 bits/word. The middle distribution for all three panels is that of the entropies of the original texts, H. For this quantity there is also language dependence, with English and Chinese

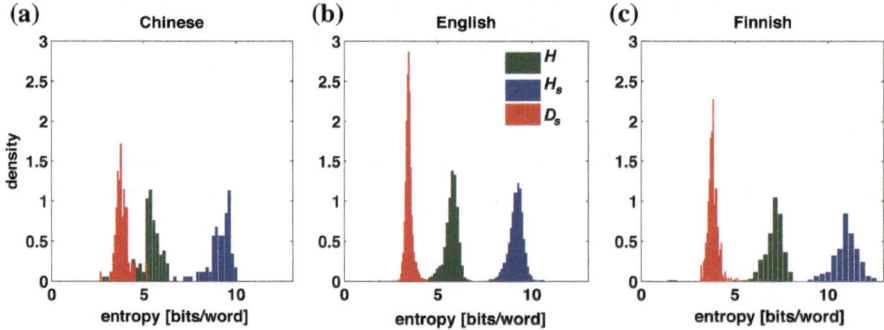

Fig. 1 Entropy distributions for corpora belonging to three languages. Each panel shows the distribution of the entropy of the random texts lacking linguistic structure (*blue*); that of the original texts (*green*); and that of the relative entropy (*red*). The three languages: Chinese, English, and Finnish, were chosen because they had the largest corpora in three different linguistic families. In panels A, B, and C, the random texts were obtained by randomly shuffling the words in the original ones. In panels D, E, and F, the random texts were generated using the words frequencies in the original texts. Adapted from [38]

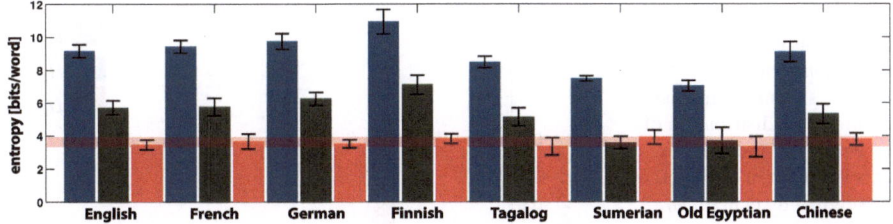

Fig. 2 Entropy of eight languages belonging to five linguistic families and a language isolate (Indo-European: English, French, and German; Finno-Ugric: Finnish; Austronesian: Tagalog; Isolate: Sumerian; Afro-Asiatic: Old Egyptian; Sino-Tibetan: Chinese. The entropies are represented with the same colours as in Fig. 1. Adapted from [38])

having lower values than Finnish. Finally, the leftmost distribution in each panel is that of D_s. The remarkable feature that emerges from this analysis is that the distribution of D_s peaks at approximately the same value—close to 3.5 bit/word—for the three languages. Furthermore, the distribution of D_s is narrower than that of the direct entropies, suggesting that much of the observed variability in the entropy distributions is due to differences in the vocabulary structure, but overall, the measure of word ordering given by D_s is less variable over each corpus.

To verify the generality of these findings we performed a similar calculation for all the 7,077 texts in our 8 corpora. The main results are shown in Fig. 2. Because of the difference in grammar and vocabulary, the values of the two entropies H and H_s show significant variability over the different languages. For example, the entropy of the disordered texts varies from 6.7 bits/word for Old Egyptian to 10.4 bits/word for Finnish, equivalent to a difference of 55 %. Correspondingly, the entropies of the intact texts change from 3.7 bits/word to 7.1 bits/word for the same languages, which is a 91 % difference. However, the relative entropy D_s shows a remarkably consistent value over the different corpora: for Old Egyptian and Finnish the values of D_s are 3.0 and 3.3 bit/word respectively, amounting to only 11 % difference. We can define the relative variability as the standard deviation of the entropies divided by the mean entropy across languages. This quantity is 0.14 for H, 0.23 for H_s, and 0.07 for the relative entropy D_s. This fact suggests that while the overall complexity of linguistic structure depends on features specific to each language, a quantification of word ordering given by the relative entropy D_s emerges as a universal feature across languages.

This remarkable constancy of the relative entropy across several linguistic families can be interpreted with the help of simple models of language. In particular, we explored Markovian models of language consisting of just a few words where all quantities of interest could be readily computed [38]. From the analysis of these simple models it turns out that in order to keep the KL divergence constant, an increase in the entropy of the random version of the texts—which is linked to the degree of diversity in the vocabulary—needs to be accompanied by a corresponding decrease in the range of correlations. Moreover, it is shown that this patterns can also

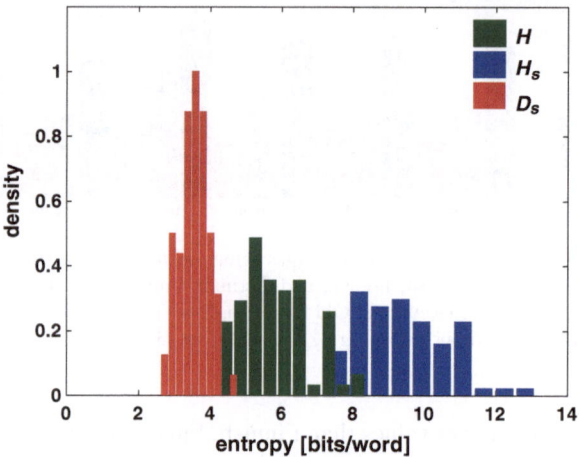

Fig. 3 Entropy distributions for a corpus of Bible translations into 75 languages from 24 linguistic families. The distribution of the entropy of the random texts in shown in *blue*, that of the original texts in *green*, and that of the relative entropy in *red*

be found in real languages, suggesting the same explanation for the constancy of the relative entropy [38].

Here we extend the previous analysis by presenting results that include several other linguistic families. We analysed a parallel corpus of translations of the Bible into 75 languages from 24 families.[2] Figure 3 shows the distribution of the entropies for all the texts in the corpus pooled together. Consistent with the results shown in Fig. 1, the distribution of the entropy of word ordering D_s is much narrower than that of the direct entropies having a mean value of 3.56 bits/word. Taking the whole corpus, the standard deviation in bits/word of H is 0.94, that of H_s 1.1 and that of D_s 0.4. In Table 1 we show the results grouped according to family, indicating as well the numbers of texts in each group.

In this first section we reviewed and extended evidence for the universal value of the relative entropy across human language. In the next section, we will discuss another relative entropy measure that, by assessing the specificity words to different contexts, can quantify and extract semantic information from language samples for which essentially no prior knowledge of the underlying linguistic structure is required.

3   An Entropy Measure of Semantic Information

In linguistic sequences both grammatical and semantic constraints affect the specific use of words at different ranges. At scales of the typical sentence length both grammatical and semantic constraints affect linguistic structure, whereas order at longer scales is mostly shaped by semantic requirements. Consequently, words that

---

[2]The parallel Bible corpus was compiled by Christos Christodoulopoulos from the Cognitive Computation Group at the University of Illinois, and downloaded from http://homepages.inf.ed.ac.uk/s0787820/bible/. Only texts that were encoded in Latin characters were used.

Table 1 Entropy values for the Bible translations grouped into 24 linguistic families

Family	Languages	H_s (bits)	H (bits)	D_s (bits)
Afro-Asiatic	5	9.51309	5.82595	3.68714
Algic	1	9.6777	6.48156	3.19614
Altaic	1	10.8493	6.52319	4.32611
Arawakan	1	10.9679	7.6054	3.36249
Austro-Asiatic	1	9.09229	5.43471	3.65759
Austronesian	7	8.67103	5.31961	3.35142
Basque	1	10.8878	7.23356	3.65423
Caribbean	1	7.37485	4.39228	2.98257
Chibchan	1	8.2331	4.97114	3.26196
Esperanto	1	9.27574	5.72075	3.55499
Creole[a]	2	7.75661	4.88113	2.87549
Equatorial	1	9.7295	5.78892	3.94058
Indo-European	25	9.65085	6.05047	3.60038
Jivaroan	3	10.968	7.13059	3.83745
Mayan	6	8.43563	4.67096	3.76467
Niger-Congo	6	10.1939	6.57799	3.61586
Nilo-Saharan	2	8.14134	5.22573	2.91561
Oto-Manguean	2	7.57349	4.40507	3.16842
Quechuan	1	10.9154	7.40662	3.50876
Sino-Tibetan	1	9.36685	6.04816	3.31869
Tucanoan	1	8.38597	4.38248	4.00349
Finno-Ugric	3	10.7777	6.83106	3.94662
Uto-Aztecan	1	9.27341	5.97364	3.29977
West Papuan	1	8.21611	4.83155	3.38456

[a]The two creole languages are Aukan and Haitian

are specific to the topics being addressed in a text show a different overall distribution compared to words that have a more structural role in language. In particular, several studies have confirmed that the words more relevant to the topics in a text tend to have an irregular distribution characterised by clustered, or bursty, patterns of occurrence [39–45]. On the contrary, function words, which are no context-specific, appear more uniformly distributed. Along similar lines we have previously reported a measure based on the entropy in the distribution of words over a text that could be used to discriminate between words belonging to different grammatical classes [46].

This insight can be incorporated into a measure of semantic information based on information theory. A measure of the information in the word distribution is based on the observation that the relevant words, or keywords, in a text are typically more dependent on the specific thematic context than non-informative words. Therefore, the specific distribution of these words can be used to distinguish statistically different parts of a text. For example, a word that only appears in one specific chapter of a book is a perfect tag for that chapter, i.e. if that word is found, it is known with certainty which chapter is being read. Despite the majority of words will have a less concentrated distribution over the text, the non-uniformities in their distribution can still be used to link them to specific contextual domains.

Consider a text of N words in length, containing K different words. The text is divided into P equal parts, of length $s = N/P$. For every word w that appears n times in the text, we can define its distribution over the text as the probability $p(w|j)$ of finding that particular word in part j ($j = 1 \ldots P$). This probability is estimated as the ratio n_j/s, where n_j is the number of occurrences of word w in part j, and is normalised as $\sum_{w=1}^{K} p(w|j) = 1$. Let us call $p(j) = 1/P$ the a priori probability that a given word w appears in part j, then the overall probability of occurrence of the word is $\sum_{j=1}^{P} p(w|j)p(j) = p(w)$, where $p(w) = n/N$. After observing an instance of word w, the probability that it comes from part j is given by $p(j|w)$, which can be computed as $p(j|w) = p(w|j)p(j)/p(w)$, or explicitly in terms of word occurrences as $p(j|w) = n_j/n$. Then, the mutual information between the sections of the text and the distribution of words is [27]:

$$M(J, W) = \sum_{w=1}^{K} p(w) \sum_{j=1}^{P} p(j|w) \log_2 \frac{p(j|w)}{p(j)}. \tag{3}$$

Since most of the words appear in the text a number of times $n \ll N$, there will be statistical fluctuations in their distribution over the partition that may induce an overestimation of the mutual information. We can correct for this bias by subtracting the mutual information computed over randomised versions of the text obtained by shuffling all the words positions. Despite it being computed over random versions of the text—where all the relationships between the words and its original contexts is lost—this quantity will not be zero in general due to the presence of statistical fluctuations. Let us call $\hat{M}(J, W)$ the mutual information estimated from one realisation of the shuffled text. Then, we can define the information in the distribution of words as $\Delta I(s) = M(J, W) - \langle \hat{M}(J, W) \rangle$, where the average is taken over an infinite number of realisations of the word shuffling. Then, using Eq. (3) and regrouping terms leads directly to the following equation for the corrected information:

$$\Delta I(s) = \sum_{w=1}^{K} p(w) \left[\langle \hat{H}(J|w) \rangle - H(J|w) \right], \tag{4}$$

where the sum is taken over the whole vocabulary of K words. Thus, Eq. (4) represents an information measure quantifying the degree of specificity of words over contextual domains characterised by the scale s. The entropy term $H(J|w)$ can be directly computed from the word counts across the P parts of the text as follows:

$$H(J|w) = -\sum_{j=1}^{P} \frac{n_j}{n} \log_2 \frac{n_j}{n}. \tag{5}$$

This quantity indicates how non-uniform is the use of word w over the text: the smaller the entropy the more non-uniform its distribution. The other entropy that appears in Eq. (4), $\langle \hat{H}(J|w) \rangle$, is similar to $H(J|w)$ with the difference that it is computed over randomly shuffled versions of the text, and averaged over an infinite number of realisations of the shuffling. This last term accounts for the fluctuations that are expected due to the finite number of occurrences of words and can be computed analytically [44]. The information given by ΔI is then a an average measure of how non-random is the distribution of words over a text.

We can gain insight into the meaning of Eq. (4) by analysing the behaviour of the two entropy quantities for an actual text. Figure 4 shows the entropies as functions of frequency computed for all the words in *The Analysis of Mind* by Bertrand Russell. Yellow dots show the entropy obtained after a random shuffling of all the words in the text and the black line is the average taken over an infinite number of random shufflings as given by the analytical estimation of $\langle \hat{H}(J|w) \rangle$ (see [44] for details). The entropy of the randomly distributed words versus frequency shows the overall trend that is expected as a consequence of statistical fluctuations in the ordering of words. High frequency words in the random text will typically have more uniform distributions than low frequency words. Finally, the entropy for the words in the original text are represented as blue dots. While the same overall trend as for the random text is observed, most of the words in the original text show values of the entropy significantly smaller than those in the randomly shuffled text for the same frequency. This is a consequence of the strong ordering constraints imposed by the linguistic structure in the original texts. Therefore, the contribution of a word to the total information is equal to this entropy difference weighted by the frequency of the word.

Figure 5 shows the information in the distribution of words for three books in English: *Opticks* by Isaac Newton, *The Analysis of Mind*, by Bertrand Russell and *On the Origin of Species*, by Charles Darwin. For small and large values of the scale parameter setting the size of the contextual domains, the value of the information approaches zero. This is because in the extremes the distribution of words in the real and random texts become identical over the partition. At intermediate values all distributions show an optimal value of the scale at which the difference in the distribution of words between the original and shuffled texts is maximised. This optimal scale represents the typical size of the partition at which the distribution of words over the text becomes most heterogeneous. For the three texts shown in Fig. 5 the values of s—in number of words—at which the information is maximal is close to 950 for *Opticks*, 750 for *The Analysis of Mind*, and 1930 for *On the Origin of Species*. The range of the optimal scales is much larger than the scope of grammatical rules and is determined by the semantic structure of the texts. An analysis done on more than 5,000 books written in English supports the conclusion that the optimal scale is related to the typical size of semantic domains over which subtopics are developed [44].

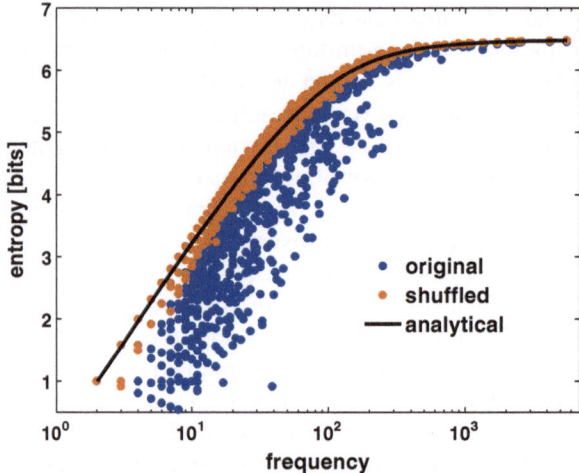

Fig. 4 Entropy of words in real and random texts. *Blue* (online) *dots* correspond to the entropy $H(J, w)$ of words in *The Analysis of Mind* (see Eq. (5)), using the scale at which the information in the distribution of words is maximal ($s = 750$ words); *orange* (online) *dots* represent the entropies computed over a randomly shuffled version of the text. The *black full line* corresponds to the analytical estimation of entropy of the random text averaged over an infinite number of the realisations of the shuffling (see [44] for details)

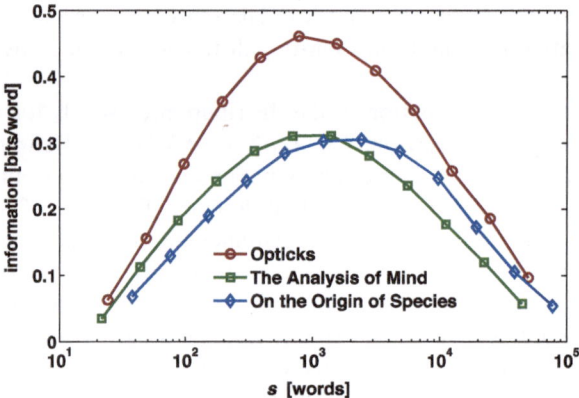

Fig. 5 Information in the distribution of words for three books in English. The *curves* represent the estimation of the information given by Eq. (4). The texts are *Opticks* by Isaac Newton, *The Analysis of Mind*, by Bertrand Russell, and *On the Origin of Species*, by Charles Darwin. For each text the information in bits/word is shown as a function of the scale parameter s determining the size of contextual domains

Table 2 List of most informative words for three texts

Opticks	Origin of species	The analysis of mind
Rings	On	Image
Colours	Species	Memory
Prism	Varieties	Images
Paper	Hybrids	Word
The	Forms	Belief
Red	Islands	Words
Light	Of	Desire
I	Will	Sensations
Rays	Selection	You
Glass	Genera	Past
Bodies	Plants	Knowledge
Colour	Seeds	Box
Image	Sterility	Content
Was	Fertility	Consciousness
Blue	Characters	Appearances
Refraction	Breeds	Movements
Water	Groups	Mnemic
Greek	Water	Feelings
Lens	The	Proposition

3.1 Information in the Distribution of Individual Words

From Eq. (4) it is apparent that the total information is a sum of contributions from individual words. Each word can then be assigned an information value equal to its weight in the sum, as $\Delta I_w(s) = p(w)\left[\langle \hat{H}(J|w)\rangle - H(J|w)\right]$. This means that the information associated with individual words depends both on their frequency and on the difference of the entropies computed on the real text and on a random version of it. In order to be informative, a word must be frequent and at the same time have a heterogeneous distribution over the text as a consequence of its specificity to its relevant semantic contexts. When words are ranked by their contribution to the overall information in a text, the top words are those more closely related to its semantic content [44]. Table 2 lists the most informative words for the three books used in Fig. 5, where the information was estimated at the optimal scale in each case. The majority of these words relate closely to the major themes in each book. The few cases of functional words that appear in the list are due to the fact that fluctuations in their distribution are greatly magnified by their high frequency. It is interesting to note that the only knowledge about the structure of the language that is incorporated into the calculation of the information is the distinction between word tokens.

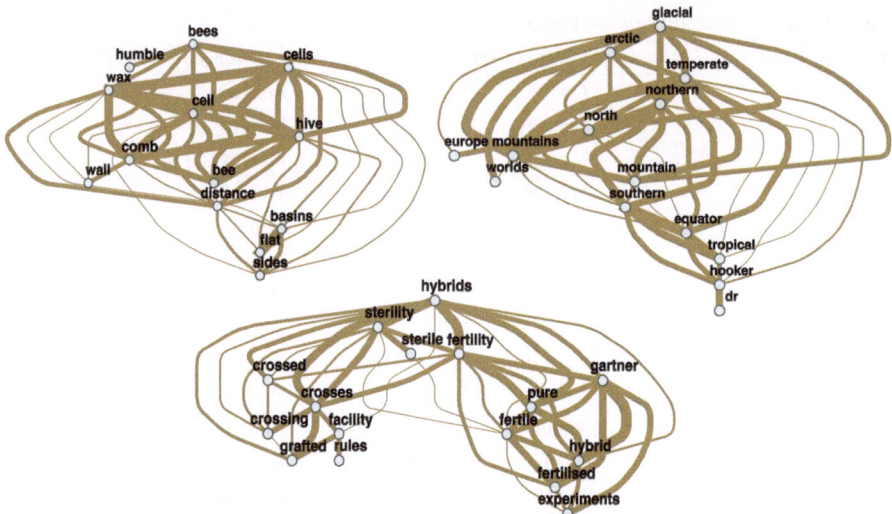

Fig. 6 Semantic networks from *On the Origin of Species*. The networks are examples obtained form the analysis of co-occurrences of the 500 most informative words without any prior knowledge about the underlying linguistic structure. The thickness of the edges indicate the strength of the connections

3.2   Extraction of Semantic Networks

Once the most informative words are identified, it is possible to study their co-occurrences in a systematic way with the aim of identifying groups of related words that tend to appear in similar contexts. Words for which their co-occurrence is statistically robust will most likely present semantic relationships. Both in [47, 48] the information-based keyword extraction was complemented with a word-space analysis in order to capture the structure of semantic networks. Each word was represented as a vector whose components are the normalised frequencies over the different parts of the text. The particular scale of the partitions was chosen as that maximising the information given by Eq. (4). Every word w can be represented by a vector of unit length, $\mathbf{u}_w = \mathbf{f}_w/|\mathbf{f}_w|$, where $\mathbf{f}_w$ is a vector with dimension P whose components are the frequencies of occurrence of the word w over the different parts, and $|\cdot|$ is the ℓ^2-norm. A similarity matrix $\mathbf{S}$ can then be built with elements defined as $(\mathbf{S})_{ww'} = \mathbf{u}_w \cdot \mathbf{u}_{w'}$. Since the vectors $\mathbf{u}_w$ are normalised, the scalar product $\mathbf{u}_w \cdot \mathbf{u}_{w'}$ equals the cosine of the angle between the vectors. Therefore, pairs of words that have a similar profile of use over the partition will be represented by normalised vectors sustaining a small angle, and consequently a scalar product close to unity. Given that all the vector components are positive, the minimum possible value for the matrix coefficients is zero. However, the similarity coefficient between two words would be zero only when these words are never simultaneously used in any of the P parts of the text, which is statistically rare. Therefore, all the elements of the similarity matrix are generally non-zero for all pairs of words, with higher values for more similar frequency profiles. An appropriate threshold applied to the similarity matrix—keeping the strongest connected word pairs—renders a set of tightly connected semantic networks. As an example, Fig. 6 shows three semantic networks obtained from *On the Origin of Species* by Charles Darwin, where words in each network clearly relate to each other. The analysis was done using the first 500 most informative words and only the 1 % strongest connections were kept. To verify that all these connections were statistically significant, the similarity matrix was evaluated repeatedly on randomised frequency vectors, in which the indices corresponding to the partitions of the text were shuffled independently for each vector. The procedure allowed us to compute a p-value for every connection, given by the fraction of times the connection of the randomised realisation was equal to or higher than that from the original text. Only links that were statistically significant ($p < 0.01$) were kept.

4 Conclusions

We have presented a summary of recent progress on the characterisation of the structure of language by means of entropy measures. We first addressed the controversial question of whether there are universal statistical patterns in word ordering. We showed that while estimations of the direct entropy over different languages yield values that strongly depend on the particular language, a measure of relative entropy that specifically quantifies the degree of word ordering presents an almost constant value over a wide range of linguistic families. This relative entropy measure can be shown to be equivalent to the Kullback–Leibler divergence between a linguistic sequence and a disordered version of it. We have also discussed some steps towards the interpretation and implications of this constancy. By using simple models and analysis of real languages, we showed that the constancy of the relative entropy requires an interplay between the diversity of their vocabularies and the extent of correlations. The degree of universality shown in this feature suggests that languages from a wide range of families have evolved under the precise constraint of keeping the relative entropy constant.

In the second part of this communication we discussed an information measure that quantifies how informative is the distribution of words in a text over the different sections of a partition of it. This quantity also relies on the assessment of the balance between order and disorder in language. One interesting insight provided by this measure is the presence of a scale in language which seems to be related to the typical lengths—in words—over which specific topics are developed in language. Moreover, since the information is additive over the words in a text it is possible to ascribe an individual information value to each word defined as its weight in the overall sum, which allows the ranking of words by their frequency distribution. The words that contribute the most to the information turn out to be the ones most closely related to the semantic content of the texts, thus providing an method for automatic keyword extraction that requires essentially no knowledge of the underlying linguistic structure of the texts.

The methods described in this paper show that a careful assessment of the balance between order and disorder in linguistic sequences with methods from statistical physics and information theory can offer significant clues into the structure of language. Still, many questions remain open, as for example what are the actual universal mechanisms that constrain the evolution of language along trajectories of constant Kullback–Leibler divergence, or whether there are further insights into the link between *meaning* and statistics.

References

1. Von Frisch, K.: The Dance Language and Orientation of Bees. Harvard University Press, Cambridge (1967)
2. Ouattara, K., Lemasson, A., Zuberbühler, K.: Proc. Natl. Acad. Sci. **106**(51), 22026 (2009)
3. Kershenbaum, A., Bowles, A.E., Freeberg, T.M., Jin, D.Z., Lameira, A.R., Bohn, K.: Proc. R. Soc. B: Biol. Sci. **281**(1792), 20141370 (2014)
4. Maynard Smith, J., Szathmáry, E.: The Major Transitions in Evolution. W.H. Freeman Spektrum, Oxford, New York (1995)
5. Deacon, T.W.: The Symbolic Species: The Co-Evolution of the Brain and Language. WW Norton & Co, New York (1997)
6. Zanette, D.H.: arXiv preprint arXiv:1412.3336 (2014)
7. Greenberg, J.H.: Universals of Language. M.I.T. Press, Cambridge (1963)
8. Chomsky, N.: Aspects of the Theory of Syntax. M.I.T. Press, Cambridge (1965)
9. Hawkins, J.A.: Word Order Universals. Academic Press, New York (1983)
10. Hawkins, J.A.: Efficiency and Complexity in Grammars. Oxford University Press, Oxford (2004)
11. Evans, N., Levinson, S.C.: Behav. Brain Sci. **32**(05), 429 (2009)
12. Dunn, M., Greenhill, S.J., Levinson, S.C., Gray, R.D.: Nature **473**(7345), 79 (2011)
13. Gell-Mann, M., Ruhlen, M.: Proc. Natl. Acad. Sci. **108**(42), 17290. doi:10.1073/pnas. 1113716108. http://www.pnas.org/content/108/42/17290.abstract (2011)
14. Zipf, G.K.: The Psycho-Biology of Language; An Introduction to Dynamic Philology. Houghton Mifflin Company, Boston (1935)
15. Heaps, H.S.: Information Retrieval, Computational and Theoretical Aspects. Academic Press, New York (1978)
16. Montemurro, M.A.: Phys. A **300**(3–4), 567 (2001)
17. Montemurro, M.A., Zanette, D.H.: Glottometrics **4**, 87 (2002)
18. Zanette, D.H., Montemurro, M.A.: J. Quant. Linguist. **12**, 29 (2005)
19. Altmann, E.G., Gerlach, M.: arXiv preprint arXiv:1502.03296 (2015)
20. Schenkel, A., Zhang, J., Zhang, Y.C.: Fractals-Complex Geom. Patterns Scaling Nat. Soc. **1**(1), 47 (1993)
21. Ebeling, W., Neiman, A.: Phys. A **215**(3), 233 (1995)
22. Montemurro, M.A., Pury, P.: Fractals **10**, 451 (2002)
23. Alvarez-Lacalle, E., Dorow, B., Eckmann, J.P., Moses, E.: Proc. Natl. Acad. Sci. U.S.A. **103**(21), 7956 (2006)
24. Altmann, E.G., Cristadoro, G., Esposti, M.D.: Proc. Natl. Acad. Sci. U.S.A. **109**(29), 11582 (2012)
25. Shannon, C.E.: Bell Syst. Tech. J. **30**(1), 50 (1951)
26. Cover, T.M., King, R.C.: IEEE Trans. Inf. Theory **24**(4), 413 (1978)

27. Cover, T.M., Thomas, J.A.: Elements of Information Theory, 2nd edn. Wiley-Interscience, Hoboken (2006)
28. Lempel, A., Ziv, J.: IEEE Trans. Inf. Theory **22**(1), 75 (1976)
29. Ziv, J., Lempel, A.: EEE Trans. Inf. Theory **23**(3), 337 (1977)
30. Ziv, J., Lempel, A.: IEEE Trans. Inf. Theory **24**(5), 530 (1978)
31. Wyner, A.D., Ziv, J.: IEEE Trans. Inf. Theory **35**(6), 1250 (1989)
32. Kontoyiannis, I., Suhov, Y.M.: In: IEEE International Symposium on Information Theory, pp. 194–194. IEEE (1994)
33. Schurmann, T., Grassberger, P.: Chaos **6**(3), 414 (1996)
34. Kontoyiannis, I., Algoet, P.H., Suhov, Y.M., Wyner, A.J.: IEEE Trans. Inf. Theory **44**(3), 1319 (1998)
35. Puglisi, A., Benedetto, D., Caglioti, E., Loreto, V., Vulpiani, A.: Phys. D: Nonlinear Phenom. **180**(1), 92 (2003)
36. Gao, Y., Kontoyiannis, I., Bienenstock, E.: Entropy **10**(2), 71 (2008)
37. Lesne, A., Blanc, J.L., Pezard, L.: Phys. Rev. E **79**(4), 046208 (2009)
38. Montemurro, M.A., Zanette, D.H.: PLOS ONE **6**(5), e19875 (2011)
39. Harter, S.P.: J. Am. Soc. Inf. Sci. **26**(4), 197 (1975)
40. Church, K.W., Gale, W.A.: Nat. Lang. Eng. **1**(2), 163 (1995)
41. Ortuño, M., Carpena, P., Bernaola-Galvan, P., Munoz, E., Somoza, A.M.: Europhys. Lett. **57**(5), 759 (2002)
42. Herrera, J.P., Pury, P.A.: Eur. Phys. J. B - Condens. Matter Complex Syst. **63**(1), 135 (2008)
43. Altmann, E.G., Pierrehumbert, J.B., Motter, A.E.: PLOS ONE **4**(11), e7678 (2009)
44. Montemurro, M.A., Zanette, D.H.: Adv. Complex Syst. **13**(2), 135 (2010)
45. Carretero-Campos, C., Montemurro, M.A., Bernaola-Galván, P., Coronado, A.V., Carpena, P.: In: *Proceedings of the European Conference on Complex Systems*, pp. 241–249. Springer (2013)
46. Montemurro, M.A., Zanette, D.H.: Adv. Complex Syst. **5**(1), 7 (2002)
47. Montemurro, M.A., Zanette, D.H.: PLOS ONE **8**(6), e66344 (2013)
48. Montemurro, M.A.: Cortex **55**, 5 (2014)

Symmetry and Universality in Language Change

Richard A. Blythe

Abstract We investigate mechanisms for language change within a framework where an unconventional signal for a meaning is first innovated, and then subsequently propagated through a speech community to replace the existing convention. We appeal to the notion of universality as it applies to complex interacting systems in the physical sciences and which establishes a link between generic ('universal') patterns at the macroscopic scale and symmetries at the microscopic scale. By relating the presence and absence of specific symmetries to fundamentally distinct mechanisms for language change at the level of individual speakers and speech acts, we are able to draw conclusions about which of these underlying mechanisms are most likely to be responsible for the changes that actually occur. Since these mechanisms are typically believed to be common to all speakers in all speech communities, this provides a means to relate universals in individual behaviour to language universals.

1 Three Notions of Universality in Language Change

Language is a system of behaviour that is acquired by social learning, that is, by learning from other members of a social group as opposed to a process of individual exploration [1]. On the face of it, the social interactions where a linguistic behaviour is transmitted from on individual to another are highly specific. Each interaction could depend on the the goals of the participants in the interaction, their own individual history of usage, the relative social standing of the individuals involved, to name just three factors that have been discussed in the literature [2–4]. Nevertheless, when one looks at the system that arises from these repeated social interactions, common patterns emerge.

R.A. Blythe (✉)
School of Physics and Astronomy, SUPA, University of Edinburgh,
Peter Guthrie Tait Road, Edinburgh EH9 3FD, UK
e-mail: R.A.Blythe@ed.ac.uk

© Springer International Publishing Switzerland 2016
M. Degli Esposti et al. (eds.), *Creativity and Universality in Language*,
Lecture Notes in Morphogenesis, DOI 10.1007/978-3-319-24403-7_4

   Some of these patterns relate to the structure of language itself. For example, typological surveys show that although six different orderings of the subject (S), verb (V) and object (O) are possible, two particular orderings (SOV and SVO) are much more common than any of the others (see Feature 81 A in [5]). Other patterns relate to how languages change over time, in particular those cases where one conventional signal for a meaning is replaced by another [3]. A number of these linguistic patterns are surveyed in [6]. These include the 'male lag', which relates to the common observation that when a change is in progress, it is the females who lead the change (i.e. are less likely to be users of the outgoing convention). Meanwhile, when partitioning language users by age, rather than gender, one typically encounters an 'adolescent peak', whereby the age group leading the change is not the very youngest, but the adolescents. Finally, the frequency of the new convention as a function of time tends to follow an S-curve, that is starting slowly, then accelerating, before tailing off as the old convention is eliminated. Indeed, this pattern is seen not only in language change, but also in other types of cultural change, such as the adoption of a technological innovation [7]. All of the phenomena described in this paragraph might be described as *universal*, in the sense that they have been observed in different social groups at different times, and in some cases even across more than one type of cultural behaviour.

   This however is not the only possible notion of universality that relates to language change (or cultural evolution more generally). The linguistic behaviour that is displayed and transmitted in social interactions is determined to some extent by the cognitive and physical apparatus possessed by the interacting agents. For example, in the case of word order, it is possible that sentences that have the subject first are easier for humans to process than other types of sentence, which would be expected to lead to those subject first sentences being more common across the world's languages. A variety of such linguistic principles have been proposed: see e.g. [8] for a discussion in a psychological context. Likewise, articulatory or auditory constraints may cause certain vocalisations to be more easily produced or understood than others [9]. The crucial point is that these constraints are assumed to be common to all language users, no matter which social group they belong to: in this sense (and one that is distinct to the above) these abilities are *universal*.

   It is natural to expect some sort of link between these two types of universals: that is, to propose that the origin of universal patterns of cultural evolution lies in the universal constraints that underpin the social interactions and social learning. What is unclear is whether the relationship is simple and transparent. In this case, every phenomenon that is seen at the macroscale would be directly observable in individual interactions. On the other hand, the relationship might be rather more complex, arising from multiple biases and the fact that the behaviour has been acquired and reproduced multiple times. Experimental work provides evidence in favour of both positions (e.g. [10, 11]), which is perhaps not surprising since they are not mutually exclusive.

   One tool that is becoming increasingly widely used to understand the link between universals at the individual and population level is mathematical modelling of complex interacting agent systems [12–14]. Here a great deal of intuition is drawn from the experience of modelling physical systems of interacting particles (atoms and molecules) that collectively form macroscopic structured materials (for example, metals). In this context, one encounters a notion of *universality* that is again distinct to the two cited above. Roughly speaking, this notion pertains to the link between individual-level and collective behaviour, and in this work we shall draw inspiration from it to understand how the way in which a language change propagates can be related to individual behaviour.

It is instructive to discuss briefly a concrete example of universality in condensed matter physics to elucidate our approach. Magnetic materials are characterised by a *Curie temperature*, below which they exhibit permanent magnetism [15]. For iron, the Curie temperature is 770°C, which is why an iron bar serves as a good choice for a bar magnet in a child's chemistry set.[1] At a distance ΔT below the Curie temperature, the strength of the magnet increases as a power-law $(\Delta T)^\beta$ (at least in the range where ΔT is small) [16]. It is this exponent β that is universal: it has the same numerical value for a wide variety of magnets with different microscopic structures. For example, model magnets whose component parts interact with different strengths or different ranges, or have different spatial arrangements, all have the same power-law exponent β [16].

We can now state more precisely what is meant by universality in this context. It applies when some macroscopic phenomenon is observed independently of the details of the interactions between the component parts *as long as* these interactions are consistent with a certain set of general principles. In condensed matter physics, these principles relate to the *symmetry* of the system [16]. In the example of the magnet, the relevant symmetry property is that the interactions are unaffected if one exchanges all north and south poles of the microscopic magnets that collectively form the macroscopic magnet.

In the remainder of this article, we will see how similar ideas relating to symmetry in linguistic interactions between speakers can be used both to predict the emergent dynamics of language change, and to categorise different theories for the factors that may influence individual behaviour. As we will see, various types of asymmetry are possible, and each corresponds to a characteristic pattern of language change, only some of which are consistent with the universal S-curve of language change mentioned above (and discussed in further detail below). While the main results outlined here were established in the context of a specific model in Ref. [17], we offer here a much broader perspective than was achieved in this earlier work. In particular, we present some new general results that apply to a wide range of models that respect the relevant symmetries while differing in detail. As such, they underline the utility of considerations based on symmetry as a means to understand the behaviour of complex interacting systems outside the physical sciences.

---

[1] One may ask what a magnet is doing in a 'chemistry set', given that magnetism is physics, but this is beyond the scope of this article.

2 Asymmetry in Language Change: The Universal S-Curve

We begin by stating more precisely the properties of the universal S-curve of language change, and highlight why this points towards some underlying asymmetry in the system. Throughout this work, we will always have in mind the the case of a language change in which the conventional signal for a specific meaning is replaced over time with a new signal. Specific examples include the marking of the future tense in Brazilian Portuguese [18], negation in French [19] and the word used by English speakers in Canada to refer to the item of furniture that I (and the people I typically interact with[2]) call a 'sofa' [20]. In each of these cases, the frequency that the incoming variant ('couch') is used follows an S-curve trajectory: the rate of growth initially accelerates until both incoming and outgoing variant are widely used, after which the rate of growth decelerates as the incoming variant becomes established as a convention (i.e. a variant that is used by a large majority of speakers). The empirical data for Canadian furniture terms is shown in Fig. 1, where the rise of 'couch' has been fit to an idealised logistic S-curve that has been discussed widely in the historical and sociolinguistics literature (see e.g. [3, 21–24]).

The case of Canadian furniture terms is an interesting one as multiple variants are in simultaneous competition. Among the older speakers, 'sofa' and 'couch' are the most used low frequency variants, and indeed given this information, one might expect 'sofa' to be the term most likely to displace the existing convention ('chesterfield'). This shows that there are (at least) two factors that determine the dynamics of one of the variants over time: its initial frequency f_i, and its rate of growth, s_i. In the absence of competition, the frequency grows as $x_i(t) = f_i e^{s_i t}$. The effect of competition is included by ensuring that all the frequencies sum to 1: $\sum_i x_i(t) = 1$. Then,

$$x_i(t) = \frac{f_i e^{s_i t}}{\sum_j f_j e^{s_j t}} \ . \tag{1}$$

Depending on the initial frequencies and growth rates, one can arrive at a variety of different shapes of curve, as shown in Fig. 1. The key point is that the variant with the largest growth rate (s_i) will eventually saturate to $x_i(t) = 1$ and, if it starts at low frequency, will typically follow the characteristic S-shaped curve.

Another way to understand the S-curve—and in particular its symmetry properties—is to take a dynamical systems theory view. The rate of change of the variant frequency when it is close to 0 % and 100 % is sufficiently small that it can be idealised to zero. This implies that these are fixed points of the dynamics. However, the initial state is an unstable fixed point (repulsive) while the final state is stable (attractive). Thus there is an asymmetry in the stability of these two fixed points, which in turn points towards some underlying asymmetry in the system of linguistically-interacting agents. As we will see in the following, there are a number of ways in which this asymmetry may be generated: however, not all of them are equivalent in terms of the language change trajectories that arise.

---

[2]An exception is my young child, who has elected for 'couch'.

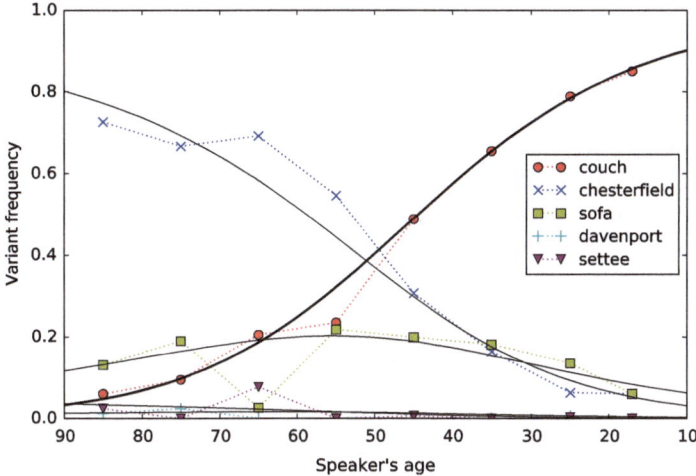

Fig. 1 Variation in frequency of different furniture terms in Canadian English. Points connected with dotted lines correspond to empirical data from [20]. This was an apparent-time study, meaning that speakers of different ages were surveyed. In this framework, older speakers are assumed to be representative of typical behaviour at an appropriate point in the past. Hence plotting the data as a function of decreasing age gives an estimate of the real-time change trajectory for these data. Solid lines are fits to the functions (1); the thickest line is that with the largest growth rate, and is the ultimately winning variant that follows the S-curve

We emphasise that our paradigm throughout this work is the case where an existing convention is being replaced by an innovative variant signal for the same meaning. The innovation process itself is not directly modelled; rather, it is implicit in the initial condition, which will be a very low (but nonzero) frequency for the innovation.

3 Language Change with No Asymmetry

For orientation, we ask the following question: What would language change look like if there are no asymmetries at all? This is a very strong requirement. First, every member of a speech community must behave identically. Every group of speakers that interacts—be this in pairs, triads or larger units—must interact with the same frequency, and each speaker must react in the same way to the behaviour of the speakers they interact with. They must also give no preference to any of the variants (e.g. different words for 'sofa') over any other that they are exposed to. This already shows that there are at least three ways to generate asymmetry, and we shall consider them all below.

Suppose the innovation (incoming variant) is used in the speech community with a frequency x. Here, by *frequency*, we mean a number between 0 and 1 which corresponds to the fraction of utterances where a specific meaning is being expressed in which the innovative signal is used. By implication, we have that the convention has a frequency $1 - x$.

If no asymmetries are allowed, then all speakers can do is produce each variant in proportion to what they have heard. This means that the expected rate of change of any variant is zero: that is, the innovation frequency x is just as likely to go up as to go down in any time interval.

We can demonstrate this result by appealing to a fairly general mathematical model (and one that we will modify in later sections to explore the link between symmetry and the resulting language change process). Let G_{ij} be the probability that agents i and j interact in a time interval lasting δt. The frequency of the innovation experienced by speaker i over this time interval is

$$\tilde{x}_i = \frac{\sum_{j \neq i} G_{ij} x_j}{\sum_{j \neq i} G_{ij}} . \tag{2}$$

The quantity appearing in the denominator here $G_i = \sum_{j \neq i} G_{ij}$ is the total probability that agent i interacts with another speaker in the time interval δt.

Now if speaker i participates in an interaction in this time interval (an event that occurs with probability G_i), then we suppose that it updates its usage frequency to equal some average of its existing value x_i and the frequency $\tilde{x}_i$ observed in the interaction. Otherwise, if it does not interact (probability $1 - G_i$), the usage frequency remains unchanged. That is, the mean value of a speaker i's usage frequency after an interaction, x_i', is

$$x_i' = G_i[\alpha x_i + (1 - \alpha)\tilde{x}_i] + (1 - G_i)x_i = x_i + G_i(1 - \alpha)(\tilde{x}_i - x_i) \tag{3}$$

where α is a number between 0 and 1 that specifies how resistant a speaker is to change. In the case $\alpha = 0$, a speaker immediately accommodates to the usage frequency of its interlocutors; in the case $\alpha = 1$ it never changes. Since there are no asymmetries, α is the same for all speakers, and we avoid the pathological (and uninteresting) case of no change, $\alpha = 1$.

We can now work out what the overall frequency of the innovation is after all speakers have updated their individual frequencies. We find

$$x' = \frac{1}{N} \sum_i x_i' = \frac{1}{N} \sum_i [x_i + G_i(1 - \alpha)(\tilde{x}_i - x_i)] = x + \frac{1 - \alpha}{N} \sum_i G_i(\tilde{x}_i - x_i) \tag{4}$$

where N is the number of speakers. Notices that strictly speaking what we have calculated here is the *mean* frequency of the innovation in the population, where the average is over all possible interactions that might happen in the time interval δt. For large speech communities, we are justified in ignoring fluctuations which are expected to be of order $1/\sqrt{N}$ (unless we find that the expected changes in x are themselves of similarly small magnitude).

Now symmetry demands that $G_{ij} = G_{ji}$—the frequency that i interacts with j equals the frequency that j interacts with i. (Actually, this will always be true, although the response to the interaction need not be symmetric, as discussed below). This has the following important consequence:

$$\sum_i G_i \tilde{x}_i = \sum_i \sum_{j \neq i} G_{ij} x_j = \sum_j x_j \sum_{i \neq j} G_{ij} = \sum_j x_j \sum_{i \neq j} G_{ji} = \sum_j G_j x_j = \sum_i G_i x_i .$$

(5)

Using this in (4), we find

$$x' = x + \frac{1-\alpha}{N} \left[\sum_i G_i \tilde{x}_i - \sum_i G_i x_i \right] = x + \frac{1-\alpha}{N} \left[\sum_i G_i x_i - \sum_i G_i x_i \right] = x .$$

(6)

This shows that the expected frequency of the innovation in the speech community, x', at the end of a time interval lasting δt is the same as its value, x, at the start of that time interval. In other words, variant frequencies do not change on average.

At this point, it is necessary to return to the observation that there are fluctuations of order $1/\sqrt{N}$ around this average change in frequency. That is, in any real system we will expect to see small changes in variant frequencies from one time step to the next due to fluctuations in the identities of the speakers who interact, and their response to the interaction. However, the symmetries inherent in these interactions imply that the probabilities probabilities of upward and downward fluctuations are the same. Consequently, the small fluctuations in variant frequencies are undirected. Given enough time, it is possible for one of the variants to be eliminated by chance, at which point it will not (at least in the class of models under consideration here) be reinvented. The canonical mathematical model for this random processes with these characteristics is genetic drift that was introduced mathematically in the 1930s [25, 26]. Typical trajectories of change generated by genetic drift are shown in Fig. 2, and can be seen to differ significantly from the directed S-curve of Fig. 1. In particular, both fixed points (at $x = 0$ and $x = 1$) are stable, as one would expect if there is no underlying asymmetry.

For what follows, it is perhaps worth emphasising the symmetries assumed in this analysis. First, all possible variants are considered equivalent. Further, all speakers are equivalent in terms of their propensity to change (all have the same α value). Dyads are symmetric: when speaker i interacts with speaker j, speaker j interacts with speaker i (and they interact in the same way). More subtly, we assumed that a speaker's updated usage frequency would be some linear combination of its existing frequency and those if its interlocutors. As we will see below, nonlinear functions correspond to distinguishing between variants by their usage frequencies. A fully symmetric model would preclude giving a higher (or lower) weight to a more frequent variant. We explore the effect of relaxing each of these symmetries in the following sections.

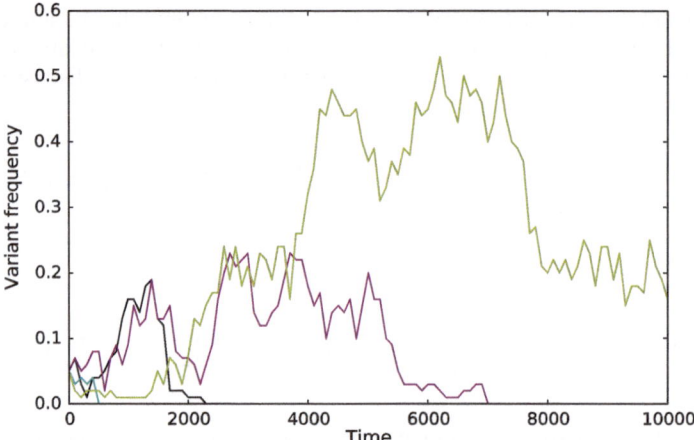

Fig. 2 Four change trajectories generated by genetic drift with an initial frequency of the innovation of 0.05. In each case, strong upward and downward fluctuations are observed, and no directed S-curve change trajectory (similar to that seen in Fig. 1) is seen. Three of the innovations go extinct (one rather quickly); one is still present at the end of the time period shown

4 Asymmetry in Interaction Frequencies

A fully symmetric model would have all interaction frequencies G_{ij} equal. Note, however, that we did not make this assumption in the previous section. Therefore, the main conclusion—that a variant's frequency exhibits undirected fluctuations—should hold for arbitrary variation in G_{ij} between pairs of speakers. It turns out that this is indeed the case, as was shown quite generally for a wide range of evolutionary processes (including cultural evolution) on complex network structures [27]. In fact, this insensitivity to network structure in the dynamics goes much deeper, in that the fluctuations in the usage frequency x_v at the community level depend only on the size of the speech community, and not on the details of who speaks to whom and how often [27, 28]. This finding is reminiscent of the concept of universality as it applies to magnets, where the spatial arrangement of atoms in the solid did not affect its overall magnetic properties (see Sect. 1 above).

It would be somewhat unreasonable to expect every member of a speech community to have the same number of interlocutors, and to interact with each of them with exactly equal frequency. Consequently, the fully symmetric theory of the previous section has never (to my knowledge) been advanced as a linguistic theory for language change. However, the extension to the case where interaction frequencies can vary and speakers adopt a some average of their own and their interlocutor's frequencies for future interactions *has* been advanced as the linguistic theory of *determinism*, at least as it applied to new-dialect formation under certain circumstances [29]. The psychological basis for this theory is accommodation, a process whereby speakers align themselves with their interlocutors for various reasons, for example, to increase their chance of being understood [3, 30]. If one looks infinitely far into the future, the outcome of a genetic drift process is quantitatively consistent with the fate of New Zealand English [31]. However, this explanation relies on fluctuations to reach the state where all agents have adopted the innovation. Recall from the previous section that the magnitude of these fluctuations decreases with the speech community size. The work of [27, 31] concludes that the timescale of change in such a theory increases with the speech community size in a way that is inconsistent with the rapid pace of language change seen in the example of New Zealand English, where the speech community was large. In order to see a more rapid change, or to see a directed change, a more powerful asymmetry is needed.

5 Asymmetry in Social Attitudes

One way to introduce further asymmetry is if speakers have different attitudes towards each others' behaviour. In particular, in an interaction between speakers i and j, there is no particular reason why speaker i should give the same weight to speaker j's utterances as the other way round. The question now is whether this asymmetry can generate the sustained directed growth of an innovation.

For this to be possible, there must be some correlation between this asymmetry and the set of speakers who initially use the innovation. To understand why, consider the opposite case where there are no correlations between the influence that a speaker has and whether they initially use the innovation. The average influence of speakers who use the innovation is then, by definition, equal to the average influence of speakers who do not. This averaging out of influence then restores the symmetry between the variants, and thus one would not expect directed changes to arise.

In models of innovation diffusion, some relationship between innovativeness and social influence is typically assumed (albeit with varying degrees of explicitness). For example, Rogers [7] refers to a group of 'innovators' who have influence over an 'early majority' who, in turn, have influence of a 'late majority' and so on. In the sociolinguistics literature, there has been some discussion of social networks, focussing on the role that strong ties between individuals might play as a mechanism to preserve social norms, and how the number and quality of relationships between in governing how linguistic variation propagates (see e.g. [32, 33]). Meanwhile, Labov [4] and Rogers [7] further emphasise the important role played by specific individuals who have influence over other members of a social group when it comes to propagating an innovation. These all imply some sort of asymmetry in social influence.

What we have found when incorporating social asymmetry into a model of language change is that an innovation which is initially used within a small group of influential users can grow in frequency over a sustained period. However, the shape of the adoption curve depends somewhat on the details of how the social network is configured [17]. This one can see by asking how the average usage frequency of the innovation changes in the presence of social asymmetry.

To this end, let us return to the mathematical model of Sect. 3, and generalise it to the case of asymmetric interactions. This we shall do by redefining the quantity G_{ij}. Previously, this was the probability that agents i and j interact in a time interval of length δt. We now take it to be equal to the probability that agents i and j interact in this time interval *and* that agent i modifies its usage frequency in response to agent j's utterances. Clearly we no longer require $G_{ij} = G_{ji}$. If agent i seeks to emulate agent j more than the other way round, we will have $G_{ij} > G_{ji}$; otherwise the converse will be true.

In terms of the mathematics, Eqs. (2)–(4) are unaffected by this redefinition. However, the relationship (5) crucially depended on the symmetry $G_{ij} = G_{ji}$. This time, we find instead from (4) that

$$x' - x = \frac{1-\alpha}{N} \sum_i \left[\sum_{j \neq i} G_{ij} x_j - \sum_{j \neq i} G_{ij} x_i \right] = \frac{1-\alpha}{2N} \sum_i \sum_{j \neq i} (G_{ij} - G_{ji})(x_j - x_i)$$

(7)

where we have twice used the fact that $\sum_i \sum_{j \neq i} f_{ij} = \sum_i \sum_{j \neq i} f_{ji}$ by exchanging indices and reversing the order of summation.

This expression shows that the interaction asymmetry $G_{ij} - G_{ji}$ is crucial in determining the rate of change of the usage frequency x. First of all, we can confirm our intuition that where the language behaviour (encoded here by the differences $x_j - x_i$) is uncorrelated with the interaction asymmetries, the above sum will be of order $1/\sqrt{N}$, and consequently we expect the dynamics to be similar to the case of no asymmetry (see Sect. 3).

Second, when $G_{ij} - G_{ji}$ is positive, we see that the usage frequency tends to increase if speaker j is more innovative than speaker i (i.e. if $x_j > x_i$), and it tends to decrease otherwise. This is to be expected, since we have $G_{ij} > G_{ji}$ when speaker i pays more attention to speaker j than vice versa. More significantly, this observation has implications for the shape of an adoption curve when the frequency of an innovation is small.

To see this, suppose initially that some speakers are innovators, and have $x_i = 1$, whilst the remainder of the speech community are all categorical uses of the existing convention, and have $x_i = 0$. Suppose also that the innovators exert influence over non-innovators. Then, for this initial condition we have

$$x' - x = \frac{1-\alpha}{N} \sum_{[ij]} (G_{ij} - G_{ji})$$

(8)

where here the notation $[ij]$ refers to ordered pairs i, j such that speaker j is an innovator and speaker i is not. The statement that the innovators exert influence over non-innovators implies that $G_{ij} > G_{ji}$ for all such pairs. Hence, $x' - x$ is a strictly positive quantity even with small numbers of innovators: that is, there is some positive rate of growth at low innovation frequencies. Simulation results [17] show that this initial growth can be sufficiently rapid to be inconsistent with an S-shaped adoption curve. In particular, it was shown that the initial period of slow growth could only be realised under conditions where the size of each successive group in the chain of adopters increased exponentially along the chain [17]. As far as we are aware, this does not match any known population structure.

In summary, when one relies on asymmetry in social attitudes to drive the adoption of an innovation, universality in the third (physics) sense does not apply. The details of the network of social influences matter, at least in terms of the initial shape of the adoption curve. Therefore, this type of asymmetry does not provide a robust explanation for the universal S-shaped trajectory of language change.

6 Asymmetry in the Variants

It turns out that a robust explanation for the S-curve is provided by an asymmetry in speakers' attitudes towards the linguistic behaviour itself rather than its users. To model this, we now introduce an explicit bias $f(x_j)$ into agent i's estimate of its usage frequency among its interlocutors. Specifically, we now take

$$\tilde{x}_i = \frac{\sum_{j \neq i} G_{ij} \left[x_j + f(x_j) \right]}{\sum_{j \neq i} G_{ij}} \tag{9}$$

instead of the unbiased expression (2). Whenever the bias $f(x_j)$ is positive, the frequency of the innovation is over-estimated relative to its actual value; likewise when it is negative, the frequency is under-estimated. We impose two constraints on the form of $f(x_j)$. First, we insist that it vanishes when $x_j = 0$ or when $x_j = 1$. This is to be consistent with our approach, in which the innovation process is implicit in the initial condition: if we did not have $f(0) = f(1) = 0$, the innovation would be spontaneously recreated if it goes extinct. We also insist that $0 \leq x_j + f(x_j) \leq 1$ for all x_j, so that it can be interpreted as a frequency in the same way as x_j.

If we take this variant-based asymmetry to be the sole asymmetry in the system, we will have $G_{ij} = G_{ji}$, as in Sect. 3. Using the above expression for $\tilde{x}_i$ in Eq. (4) we find that

$$x' - x = \frac{1 - \alpha}{N} \sum_i G_i f(x_i) . \tag{10}$$

We can now perform the same experiment as in the previous section, where we assign $x_i = 1$ to a group of innovators, and $x_i = 0$ to a group of conformists and ask for the initial shape of the change trajectory. Since $f(x_i) = 0$ in both cases, we find that $x' = x$, showing that with this initial condition, the frequency of the innovation can change only through a fluctuation. This is, however, not the same as the fluctuation-driven dynamics that arises when the dynamics are fully symmetric (as described in Sect. 3). There, $x' = x$ no matter what the individual usage frequencies x_i are. Here $x' = x$ only if all usage is categorical: as soon as some individuals show variable behaviour (for example, as they start to adopt the innovation), we will have $x' > x$ in the case where the bias acts in favour of the innovation (i.e. if $f(x) > 0$).

It is worth emphasising the crucial difference between the initial growth of an innovation arising from speaker-based asymmetry (Sect. 5) and the variant-based asymmetry just described. For speaker-based asymmetry, we found that with a small number of categorical innovators, the mean growth rate of the innovation is nonzero, which corresponds to a rapid initial growth. Here, for variant-based asymmetry, we found under the same conditions that the mean growth rate of the innovation vanishes, and so a slower initial growth arises. These expectations were confirmed with an explicit model in [17], which led us to hypothesise that this variant-based asymmetry is a crucial component of language change in real speech communities.

7   Asymmetry in Variant Frequencies

The foregoing does not cover all possible asymmetries that might exist in linguistic behaviour. In particular, one way in which variants could be discriminated is through their frequencies alone, without reference to any aspect of the behaviour itself or association with its users. This is actually a specific type of variant asymmetry, and as such can be modelled through an appropriate choice of the function $f(x)$ that was introduced in the previous section.

Suppose there are just two variants in competition with each other. Although we will allow the bias $f(x)$ to vary with frequency, we will do so in a way that is symmetric with respect to the variants: that is, the boost applied to a variant with some specific frequency x_0 is the same, regardless of which variant this is. In the two-variant case, the two frequencies are x and $1 - x$. The symmetry between them implies that the function $f(x)$ must satisfy the constraint

$$f(1 - x) = -f(x) . \tag{11}$$

Again, we will require that $f(0) = f(1) = 0$, so that any innovation is innovated only once, and all subsequent adoption of the imitation arises from social interactions.

The simplest functions that satisfy these requirements are $f(x) = 0$, which is the symmetric case previously considered, and $f(x) = x(1 - x)(2x - 1)$, which corresponds to boosting a variant if it is a majority variant, and suppressing it if it is in the minority. This type of frequency boosting, or regularisation, has been observed in a variety of frequency learning experiments, both in the linguistic and non-linguistic domain [11, 34]. A difficulty with this type of model is that there is a threshold problem: low frequency variants face an uphill struggle to reach a frequency of 50 %, which is needed for the regularisation bias to act in their favour. The presence of noise complicates matters. If the magnitude of any fluctuations is small, this reasoning (based primarily on deterministic considerations) continues to hold. However, when fluctuations are large there is in fact a transition into a regime where the regularisation bias is suppressed, and the usage frequencies fluctuate in the same way as in the fully symmetric case described in Sect. 3 [35].

This suggests that the main mechanism for propagating an innovation along an S-shaped adoption curve is if the function $f(x)$ is positive for all x, thereby providing a systematic bias in favour of the innovation at all frequencies. However, this raises the question of where this bias comes from. In [17], it is suggested that speaker-based asymmetry provides a means for speakers to create an association between a group of speakers and a particular linguistic behaviour. Once this happens, a variant-based asymmetry whose origins lie in speaker-based asymmetry may arise. Whether this scenario can be realised spontaneously through local interactions in an agent-based model is the subject of a current investigation [36].

Another possibility, raised in [37], is to distinguish between variants not in terms of their current usage frequencies, but according to whether they are increasing or decreasing. A 'momentum-based' bias [38] towards further increasing the frequency of a variant which has been increasing in the past could in principle propagate an innovation without appealing to a bias that is based on speaker identity. Again, the question of whether this can arise purely through local interactions between speakers is being investigated with reference to an agent-based model [39].

8 Discussion

In this short article, we have explored the various notions of universality in language change. Drawing inspiration from the relationship between symmetry and universality in physics, we have appealed to symmetry as a means to categorise theories for language change. Specifically, we identified the following sources of asymmetry in models of language change: variation in interaction frequencies alone (which corresponds to the theories of accommodation and determinism [29]); asymmetry in the degree of influence that speakers have over each other (which correspond to theories based on social network effects, propounded for example by Bloomfield [40], Labov [4], Milroy [32] and others); variation in the attitude towards different linguistic variants (which correspond to theories based on prestige and related social factors, advanced for example by Sturtevant [41], Labov [4] and enjoys some prominence among sociolinguists); and finally asymmetry that is based on the usage frequencies of variants (such as regularisation effects [34] and momentum-based explanations for change [38]).

The key message is that only some of these distinct sources of asymmetry are compatible with the widely-observed ('universal') S-shaped curve for the adoption of an innovation. We found that a robust model that generates the S-shaped curve can be achieved with a prestige-based explanation (i.e. different attitudes to particular ways of speaking) or potentially with a momentum-based explanation [37, 39]. In this work, this was determined primarily by investigating the initial rate of growth of an innovation within a fairly general mathematical framework, complementing existing studies that were based on specific simulation models.

A crucial open question that remains is the following. Appealing to symmetries is useful as it allows broad classes of explanations for language change to be excluded with reference only to qualitative features of empirical data. However, this is insufficient to identify a single theory for language change: more than one is compatible with the qualitative data. The challenge then is to distinguish between these remaining theories. One particular issue with a social prestige type explanation is how the bias towards one linguistic variant over another becomes embedded in the speech community. In [17] it was found that a majority of speakers should be positively disposed towards the innovation: how does this positive disposition itself spread through the speech community? The momentum-based theory of [37, 39] potentially side-steps this issue, since the variants are distinguished by their usage history. If different members of the speech community agree that an innovation is becoming more prevalent, they will all boost the frequency of the same variant. Whilst this is perhaps a more parsimonious theory, that is not in itself sufficient to conclude that it is the more appropriate one. Instead, some independent empirical evidence in favour of a specific explanation is needed. Even better would be to demonstrate that the favoured theory shows greater *quantitative* agreement with empirical data at both the individual and population level. The complexity of human behaviour and social interactions is such that this will be a challenging task, but one where sustained research effort would certainly be worthwhile.

Acknowledgments I thank my collaborators Gareth Baxter, Bill Croft, Simon Kirby, Alan McK-ane, Kenny Smith and Kevin Stadler with whom much of the work outlined here was done. I also thank Miriam Meyerhoff, Sali Tagliamonte and Peter Trudgill for the sociolinguistic insights that they have shared with me.

References

1. Boyd, W., Richerson, P.J.: Culture and the Evolutionary Process. University of Chicago Press, Chicago (1985)
2. Keller, R.: On Language Change: The Invisible Hand in Language. Routledge, London (1994)
3. Croft, W.: Explaining Language Change: An Evolutionary Approach. Longman, Harlow (2000)
4. Labov, W.: Principles of Lingustic Change II: Social Factors. Blackwell, Oxford (2001)
5. Dryer, M.S., Haspelmath, M. (eds.): WALS Online. Max Planck Institute for Evolutionary Anthropology, Leipzig. http://wals.info/ (2013)
6. Tagliamonte, S.A.: Variationist Sociolinguistics: Change, Observation, Interpretation. Wiley, New York (2011)
7. Rogers, E.M.: Diffusion of Innovations, 5th edn. Free Press, New York (2003)
8. Maurits, L.: Representation, information theory and basic word order. Ph.D. thesis, University of Adelaide (2011)
9. Ohala, J.: In: MacNeilage, P.F. (ed.) The Production of Speech, pp. 189–216. Springer, New York (1983)
10. Culbertson, J., Smolensky, P., Legendre, G.: Cognition **122**, 306 (2012)
11. Reali, F., Griffiths, T.L.: Cognition **111**, 317 (2009)
12. Castellano, C., Fortunato, S., Loreto, V.: Rev. Mod. Phys. **81**, 591 (2009)
13. Hruschka, D.J., Christiansen, M.H., Blythe, R.A., Croft, W., Heggarty, P., Mufwene, S.S., Pierrehumbert, J.B., Poplack, S.: Trends Cogn. Sci. **13**, 464 (2009)

14. Smith, A.D.M.: Wiley Interdiscip. Rev.: Cogn. Sci. **5**, 281 (2014)
15. Kittel, C.: Introduction to Solid State Physics, 8th edn. Wiley, Hoboken (2005)
16. Goldenfeld, N.: Lectures on Phase Transitions and the Renormalization Group. Addison-Wesley, Reading (1992)
17. Blythe, R.A., Croft, W.: Language **88**, 269 (2012)
18. Poplack, S., Malvar, E.: Probus **19**, 121 (2007)
19. Grieve-Smith, A.B.: The spread of change in french negation. Ph.D. thesis, University of New Mexico (2009)
20. Chambers, J.K.: J. Engl. Linguist. **23**, 155 (1995)
21. Greenberg, J.H., Osgood, C.E., Saporta, S.: In: Osgood, C.E., Sebeok, T.A. (eds.) Psycholinguistics: A Survey of Theory and Research Problems, pp. 146–63. Waverly, Baltimore (1954)
22. Kroch, A.S.: Lang. Var. Change **1**, 199 (1989)
23. Chambers, J.K.: In: Chambers, J.K., Trudgill, P., Schilling-Estes, N. (eds.) Handbook of Language Variation and Change, pp. 349–72. Blackwell, Oxford (2002)
24. Denison, D.: Motives for Language Change, pp. 54–70. Cambridge University Press, Cambridge (2003)
25. Fisher, R.A.: The Genetical Theory of Natural Selection. Clarendon, Oxford (1930)
26. Wright, S.: Genetics **16**, 97 (1931)
27. Baxter, G.J., Blythe, R.A., McKane, A.J.: Phys. Rev. Lett. **101**, 258701 (2008)
28. Blythe, R.A.: J. Phys. A: Math. Theor. **43**, 385003 (2010)
29. Trudgill, P.: New-Dialect Formation: The Inevitability of Colonial Englishes. Edinburgh University Press, Edinburgh (2004)
30. Giles, H., Smith, P.M.: In: Giles, H., St Clair, R.N. (eds.) Language and Social Psychology. Basil Blackwell, Oxford (1979)
31. Baxter, G.J., Blythe, R.A., Croft, W., McKane, A.J.: Lang. Var. Change **21**, 257 (2009)
32. Milroy, L.: Language and Social Networks. Blackwell, Oxford (1987)
33. Chambers, J.K.: Sociolinguistic Theory: Linguistic Variation and its Social Significance. Blackwell, Oxford (2003)
34. Hudson Kam, C.L., Newport, E.L.: Lang. Learn. Dev. **1**, 151 (2005)
35. Russell, D.I., Blythe, R.A.: Phys. Rev. Lett. **106**, 165702 (2011)
36. Blythe, R.A., Jones, A.H., Renton, J.: Spontaneous dialect formation in a population of locally aligning agents. In: S.G. Roberts, C. Cuskley, L. McCrohon, L. Barceló-Coblijn, O. Fehér, T. Verhoef (eds.) The Evolution of Language: Proceedings of the 11th International Conference (2016)
37. Stadler, K., Blythe, R.A., Smith, K., .Kirby, S.: In: Cartmill, E.A., Roberts, S., Lyn, H., Cornish, H. (ed.) The Evolution of Language: Proceedings of the 10th International Conference, pp. 525–6, (2014)
38. Gureckis, T.M., Goldstone, R.L.: Topics Cogn. Sci. **1**, 651 (2009)
39. Stadler, K., Blythe, R.A., Smith, K., Kirby, S.: Momentum in Language Change: A Model of Self-actuating S-shaped Curvess (2015) to appear in Language Dynamics and Change
40. Bloomfield, L.: Language. Holt, Rinehart and Winston, New York (1933)
41. Sturtevant, E.H.: An Introduction to Linguistic Science. Yale University Press, New Haven (1947)

Dynamics on Expanding Spaces: Modeling the Emergence of Novelties

Vittorio Loreto, Vito D.P. Servedio, Steven H. Strogatz
and Francesca Tria

Abstract Novelties are part of our daily lives. We constantly adopt new technologies, conceive new ideas, meet new people, and experiment with new situations. Occasionally, we as individual, in a complicated cognitive and sometimes fortuitous process, come up with something that is not only new to us, but to our entire society so that what is a personal novelty can turn into an innovation at a global level. Innovations occur throughout social, biological, and technological systems and, though we perceive them as a very natural ingredient of our human experience, little is known about the processes determining their emergence. Still the statistical occurrence of innovations shows striking regularities that represent a starting point to get a deeper insight in the whole phenomenology. This paper represents a small step in that direction, focusing on reviewing the scientific attempts to effectively model the emergence of the new and its regularities, with an emphasis on more recent contributions: from the plain Simon's model tracing back to the 1950s, to the newest model of Polya's urn with triggering of one novelty by another. What seems to be key in the successful modeling schemes proposed so far is the idea of looking at evolution as a path in a complex space, physical, conceptual, biological, and technological, whose structure and topology get continuously reshaped and expanded by the occurrence of the new. Mathematically, it is very interesting to look at the consequences of the interplay between the "actual" and the "possible" and this is the aim of this short review.

V. Loreto (✉) · V.D.P. Servedio
Physics Department, Sapienza University of Rome, Rome, Italy
e-mail: vittorio.loreto@roma1.infn.it

V. Loreto · F. Tria
ISI Foundation, Torino, Italy
e-mail: fratrig@gmail.com

V. Loreto
SONY-CSL, Paris, France

V.D.P. Servedio
Institute for Complex Systems (CNR-ISC), Rome, Italy
e-mail: Vito.Servedio@roma1.infn.it

S.H. Strogatz
Cornell University, Ithaca, NY, USA
e-mail: shs7@cornell.edu

© Springer International Publishing Switzerland 2016   59
M. Degli Esposti et al. (eds.), *Creativity and Universality in Language*,
Lecture Notes in Morphogenesis, DOI 10.1007/978-3-319-24403-7_5

1 Introduction

Historically, the notion of the new has always offered challenges to humankind. What is new often defies the natural tendency of humans to predict and control future events. Still, most of the decisions we take are based on our expectations about the future. The word *new* itself assumes many different meanings (see for instance [1]). We experience novelties very often in our daily lives. We meet new people, adopt new words, listen to new songs, watch a new movie, and use a new technology. Something can be new only for us or a few other people, or something can be brand new and change a paradigm or the habits of a whole population. This is a very significant phenomenon often referred to as innovation, a fundamental factor in the evolution of biological systems, human society, and technology. From this perspective, a thorough investigation and a deep understanding of the underlying mechanisms through which novelties and innovations emerge, diffuse, compete, and stabilize are a key to progress in all sectors of human activities.

Novelties and innovations share an important feature: they can be viewed as first time occurrences of something at the individual or collective level, respectively. Though a lot is known about the way novelties and innovations can possibly emerge in specific sectors, the general picture remains poorly understood theoretically and undocumented empirically. Most of the knowledge in this area is highly scattered among either highly specialized and applied environments or academic, abstract, and sometimes anecdotal publications. This paper aims at partially filling this gap by focusing on the mathematical modeling of the new and presenting a fairly incomplete review of how this problem has been tackled along with the ability of the different approaches presented to explain empirical data.

The reason why modeling what is new is difficult is related to a paradox that inference theories spell out very clearly. Inference is the branch of mathematics and statistics that deals with the problem of deriving logical conclusions from premises known or assumed to be true. A typical problem is that of estimating the probabilities of future events based on the observation of the past. It is interesting to report a passage from a review by Zabell on this subject [2]:

> This is not the problem of observing the 'impossible', that is, an event whose possibility we have considered but whose probability we judge to be 0. Rather, the problem arises when we observe an event whose existence we did not even previously suspect; this is the so-called problem of 'unanticipated knowledge'.

So the unanticipated knowledge seems to coincide with our intuitive notion of the new. What is new is by definition out of our modeling scheme and we are left with the problem of how to incorporate its occurrence in a coherent logical scheme. Another way to look at it is through the well-known dichotomy between the actual and the possible [3]. A very interesting notion here is that of the *adjacent possible* [4, 5]. Originally introduced in the framework of biology, the adjacent possible metaphor include all those things, ideas, linguistic structures, concepts, molecules, genomes, technological artifacts, etc., that are one step away from what actually exists, and hence can arise from incremental modifications and/or recombination of existing material. In Steven Johnson's words [6]:

> The strange and beautiful truth about the adjacent possible is that its boundaries grow as one explores them.

The definition of adjacent possible encodes the dichotomy between the *actual* and the *possible* [3]: the actual realization of a given phenomenon and the space of possibilities still unexplored. Figure 1 illustrates this idea with a cartoon. A walker is wandering on the nodes of a graph. The gray nodes are those already visited in the past while the white ones have never been visited. Once the walker visits a white node for the first time, another part of the graph appears that could not even be foreseen before visiting that node.

Though the creative power of the expansion into the adjacent possible is widely appreciated at an anecdotal level, still its importance in the scientific literature [5, 7–11] is, in our opinion, underestimated.

Recently, we introduced an original mathematical model of the dynamics of novelties correlated via the adjacent possible, and derived three testable, quantitative predictions from it [12]. The model predicts the statistical laws for the rate at which novelties happen (Heaps' law) and for the frequency distribution of the explored regions of the space (Zipf's law), as well as the signatures of the correlation process by which one novelty sets the stage for another. The predictions of this model were tested on four data sets of human activity: the edit events of Wikipedia pages (Fig. 2a, b), the emergence of tags in social annotation systems (Fig. 2c, d), the sequence of words in texts, and listening to new songs in online music catalogs.

By providing the first quantitative characterization of the dynamics of correlated novelties, these results provide a starting point for a deeper understanding of the adjacent possible and the different nature of triggering events (timeliness, spreading, individual versus collective properties) that are likely to be important in the investigation of biological, linguistic, cultural, and technological evolution. As a prelude to this ambitious research program, it is important to briefly summarize what is known

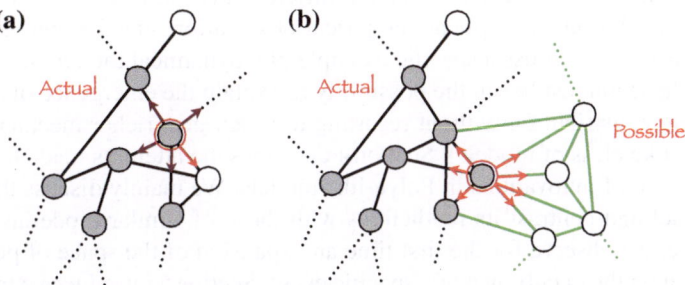

Fig. 1 Mathematical illustration of the adjacent possible in terms of a graph that conditionally expands from the situation depicted in (**a**) to that depicted in (**b**) whenever a walker visits a node for the first time (the *white node* in (**a**))

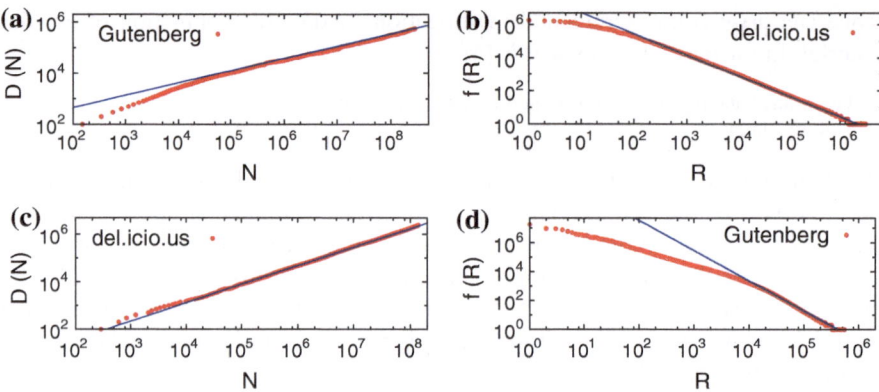

Fig. 2 Heaps' law (a–c) and Zipf's law (b–d) in real datasets. Gutenberg [13] (**a, b**), del.icio.us [14] (**c, d**) datasets. *Straight lines* in the Heaps' law plots show functions of the form $f(x) = ax^{\gamma}$, with the exponent γ equal respectively to $\gamma = 0.45$ (Gutenberg) and $\gamma = 0.78$ (del.icio.us). *Straight lines* in the Zipf's law plots show functions of the form $f(x) = ax^{-\alpha}$, where the exponent α is equal to γ^{-1} for the different γ's considered above

about the attempts made so far to model the dynamics of novelties. This is the aim of this paper.

The problem of modeling novelties is actually very old since it dates back to the work of the logician Augustus de Morgan [15], who proposed a simple way to deal with the possibility of an unknown event. For a review of this early work as well as of more recent developments, we refer to the excellent review by Sandy Zabell [2]. Here, more modestly, we merely try to summarize the main steps of how investigations in this area dealt with the problem of modeling dynamical processes on expanding spaces so far.

The outline of the paper is as follows. Section 2 deals with the class of Simon-like models where the emergence of new possibilities is ruled in a probabilistic way and the rate of innovation is constant. In this class of models, though scaling is observed for the frequencies of occurrence of the different events, there is no way, except with ad-hoc solutions, to replicate empirical observations for the innovation rates. Next, in Sect. 3 we discuss a specific example of a dynamical model on a space that shrinks, whose interest lies in the possibility to explain the emergence of scaling for the frequency distribution without resorting to a rich-gets-richer mechanism (as in the Simon-like class of models). Section 4 describes the attempts made to introduce the emergence of innovations in Polya-like models. We mainly discuss the Hoppe–Polya model and contrast its predictions with those of similar modeling schemes. In this case, we observe for the first time an expansion of the space of possibilities conditioned on the occurrence of a specific event. Section 5 introduces our modeling scheme for the adjacent possible. In this case, the expansion of the space of possibility, i.e., the adjacent possible, is triggered in a self-consistent way by the occurrence of new events. Finally, the concluding section summarizes the main conclusions and tries to highlight interesting future directions.

Throughout the paper, when the relevant statistical quantities are well described by power-laws, we shall indicate with β (minus) the exponent of the frequency distribution of tokens, with α (minus) the frequency-rank exponent and with γ the Heaps' exponent. By simple arguments reported in the Appendix, one finds that $\alpha = (\beta - 1)^{-1}$ and that $\gamma = \alpha^{-1}$ when $\alpha > 1$ while $\gamma = 1$ otherwise.

2 Simon-Like Models

The observation that the frequency distribution of words in texts written in a given language follows a fat-tailed distribution has been puzzling the scientific community since the beginning of the twentieth century [16–18] and continues to be an hot topic nowadays [19]. The search for a suitable model that could reproduce the experimental data of word frequency has also caused rather tough scientific disputes [20]. As we will briefly show in the following, almost all models based on Simon's model are not able to generate a set of tokens whose frequency distribution is represented by a power-law with an exponent $\beta < 2$ and consequently the associated Heaps' law is linear. The only exception is the model proposed by Zanette and Montemurro [21], whereas in that case the sublinear Heaps' exponent has to be recovered by data and inserted by hand without a first principle explanation.

2.1 Plain Simon's Model

One simple model able to reproduce in part the phenomenology of texts is the stochastic model devised by Simon [20]. In Simon's model a stream of tokens is generated according to the following two prescriptions: at the beginning, i.e., at time $t = 1$, only one token is present in the stream; at a generic time t a new token is added to the stream with probability p, while with complementary probability $(1 - p)$ a randomly extracted token of the stream is chosen. In this way, the tokens that appear more frequently in the stream have a higher probability to be extracted. This mechanism of favoring those elements that occur more frequently in the stream is called *rich-gets-richer*, and has become a paradigm for the generation of tokens whose frequency is distributed according to a power-law [22]. Because of its sequential nature, Simon's model is particularly suitable to describe phenomena of linguistics, such as the generation of texts, although there are still some linguistics aspects that cannot be reproduced. Above all, it is evident that the rate of addition of new tokens is constant in time (p), thus resulting in a linear growth of the available space, i.e., a linear growth of the number of different tokens $w = pt$, whereas in real texts such growth is found to be asymptotically sublinear $w = cp^{\gamma}$ with $0 < \gamma < 1$ (Heaps' law [23]). The rich-gets-richer mechanism at the basis of Simon's model, as it often happens in science [22], has been recycled in other contexts. Notably, the *preferential attach-ment* rule introduced in [24], is effectively a disguised form of the rich-gets-richer. In fact, in the Barabási–Albert model of network generation, a new node of the graph is introduced at time t by connecting it to m existing nodes chosen with probability proportional to the number of their first neighbors. This effectively corresponds to a deterministic Simon's process with the probability of extracting an old token set to $(1 - p) = 1/2$ (see Fig. 3), deterministic since at every even time step there is always a rich-gets-richer in action, while in Simon's model the entrance of an old token in the stream at any time is not certain but is conditioned to the extraction of a random number. The equivalent Simon's model so defined is characterized by a dictionary size increasing linearly in time as $D(t) = pt/m$.

The mathematical determination of the exponent characterizing the power-law distribution of the token occurrences in the Simon's model is not particularly dif-ficult [20, 22]. Here we would like to present an alternative method that involves the master equation of the process. Although this method is not rigorous, it can be straightforwardly generalized to more difficult setups. It is based on the master equa-tion and the continuous approximation. We denote with $N_{k,t}$ the number of tokens that have occurred k-times at time t and we can write its stochastic evolution in time as

$$N_{k,t+1} = N_{k,t} + (k - 1)(1 - p)\frac{N_{k-1,t}}{t} - k(1 - p)\frac{N_{k,t}}{t} + p\,\delta_{k,1}, \qquad (1)$$

i.e., the number of tokens occurring k-times at time $t + 1$ equals the number of tokens occurring k-times at time t plus the contribution we would have by extracting a token that has already occurred $(k - 1)$-times in the sequence (and this happens with probability $(1 - p)$ times the fraction of tokens occurring $(k - 1)$-times at time t), minus the contribution of extracting a token already occurring k-times (which will contribute to increase the number of tokens occurring $(k + 1)$-times), plus the specific contribution of creating a brand new token. At time t there are t tokens in

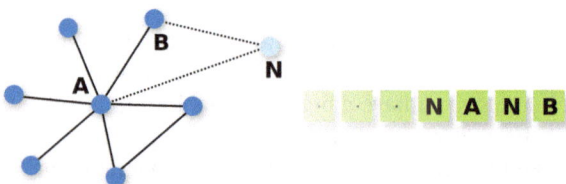

Fig. 3 Equivalence between network growth models and Simon's model. In a network growth model like that of Barabási–Albert a new node N enters the graph and attaches to both nodes A and B. This corresponds to inserting the tokens A and B and twice N in the stream of the equivalent Simon's model in which the *rich-gets-richer* mechanism happens deterministically any two time steps. Therefore, the equivalent Simon's model is obtained with the probability of extracting an old token set to $(1 - p) = 1/2$. The analogy is complete by identifying the number of occurrences of the tokens in Simon's stream with the connectivity of nodes in the graph

the stream by definition of the model, so that the probability of choosing a token occurring k-times is in fact $kN_{k,t}/t$. The continuous approximation is defined as

$$\begin{cases} N_{k,t+1} - N_{k,t} \approx \frac{\partial N_k}{\partial t} \\ kN_{k,t} - (k-1)N_{k-1,t} \approx \frac{\partial (kN_k)}{\partial k} \end{cases} \tag{2}$$

and the master equation (1) becomes

$$\frac{\partial N_k}{\partial t} = -(1-p)\frac{1}{t}\frac{\partial (kN_k)}{\partial k}. \tag{3}$$

It is verified numerically that the expression $N_{k,t}/t$ tends asymptotically to a stationary distribution q_k. Therefore, by posing $N_{k,t} = tq_k$ in Eq. (3) and by simplifying the partial derivatives, we get

$$(1-p)k\frac{dq_k}{dk} = -(2-p)q_k. \tag{4}$$

By solving the previous ordinary differential equation with standard methods, we obtain $q_k \propto k^{-1-\frac{1}{1-p}}$, i.e., a power-law distribution with exponent $\beta = 1 + \frac{1}{1-p}$ corresponding to a frequency-rank exponent $\alpha = (1-p)$. The aforementioned case of the Barabási–Albert model corresponds to setting $p = 1/2$, with a resulting network with an asymptotic distribution of node degree (number of first nearest neighbors) obeying a power-law with exponent $\beta = 3$.

2.2 Simon's Model with Time Dependent Sublinear Invention Probability

As mentioned above, the Simon's model is in some sense satisfactory but not conclusive and two main issues can be pointed out. First, with p in the range between 0 and 1, frequency-rank exponents α larger than 1 cannot be recovered although lots of idioms actually display them. Second, the dictionary, i.e., the number of different tokens (which can be interpreted as the size of the space) grows linearly in time and not sublinearly, i.e., faster than in reality. To correct both issues, a time dependent and decreasing probability $p_t = p_1 t^{\gamma-1}$ with $0 < \gamma < 1$ can be introduced in the model [21], thus assuring that the dictionary $D(t)$ would grow as t^γ, since by definition $D(t) = \int_1^t p_1 s^{\gamma-1} ds$. To see how in this case the frequency-rank distribution behaves at large t, we start from Eq. (3), set $p = p_t = p_1 t^{\gamma-1}$ and look for an asymptotic solution of the type $N_{k,t} = t^\gamma q_k$. After some algebra we come to the equation

$$(1 - p_1 t^{\gamma-1})k\frac{dq_k}{dk} = -(\gamma + 1 - p_1 t^{\gamma-1})q_k \tag{5}$$

that at large times becomes

$$k\frac{dq_k}{dk} = -(1+\gamma)q_k. \tag{6}$$

Its solution is then for large t, $q_k \propto k^{-1-\gamma}$ with a frequency-rank distribution decreasing as a power-law of exponent $1/\gamma$. At fixed large t the frequency of tokens can be written as $f_k = N_{k,t}/t \propto t^{\gamma-1}q_k$ with the time dependent factor that will be absorbed by the normalization constant.

2.3 Simon's Model with Memory

In Simon's model, there is no need to introduce an explicit time ordering, since the mechanism of extraction of old tokens from the stream is time independent. In fact it would be more appropriate to name the stream as "heap" instead. This feature is partially justified as soon as the words of texts are considered, but when the tokens have to represent other aspects of life, things may be different. In particular when tokens are identified with the songs listened to by one person or technological products on the market purchased by people, the age of tokens does matter in choosing them (e.g., nobody would buy an outdated computer).

An effect of aging can be included in Simon's model in many ways. Two of them are worth of mentioning. Dorogovtsev and Mendes [25] introduced a mechanism of aging beside the preferential attachment in the Barabási–Albert model of network growth. Their findings can be interpreted in the framework of Simon's model by the already mentioned equivalence of network growth processes and Simon's model with $p = 1/2$ (Fig. 3). In their model, a new node is attached to old nodes proportionally to their connectivity and their age. The latter is taken into account by means of a factor $\tau^{-\nu}$, with τ being the elapsed time since the token's first appearance. Interestingly, they find that for $\nu < 1$ the connectivity of nodes still obeys a power-law distribution with exponent ranging from -2, obtained as $\nu \to -\infty$, and infinity as $\nu \to 1$. When $\nu > 1$ the connectivity is described by an exponential distribution. We expect the same to hold in the general case of Simon's model for any p, where an already occurring token can be chosen with probability $(1-p)$ proportionally to its number of occurrences and proportionally to its age to the power $-\nu$. We conjecture that in the general case of $p \neq 1/2$ the exponent of the frequency distribution would be bounded above by -2, being exactly $\beta = -1 - \frac{1}{1-p}$ in the case of $\nu = 0$. It is interesting to note that the nontrivial exponential behavior is detected whenever the aging kernel $\tau^{-\nu}$ is not an integrable function of τ, i.e., when $\sum_{\tau=1}^{T}\tau^{-\nu}$ diverges at large T so that the effective size of the sampling window is infinite. The model of Dorogovtsev and Mendes still presents the same issues as Simon's model, i.e., the Heaps' law is linear and the exponent of the frequency distribution of tokens may not exceed the value -2.

A different mechanism of aging was introduced by Cattuto et al. [26, 27] to mimic the correlations and frequency of tags in social annotation systems like *del.icio.us* [14]. In that case, the choice of the old existing token in the stream was conditional to the usual probability $(1 - p)$ and to a memory kernel of the type $(r + \Delta t)^{-1}$, where $r \geq 1$ is a constant and Δt is the age of single tokens measured in time steps, i.e., tokens are no longer considered as a whole as in the Dorogovtsev–Mendes model. The usual Simon's model would correspond to a trivially constant kernel. This model results in a rather complicated token frequency distribution that resembles a power-law of exponent -2 and a frequency-rank distribution that is well described by a stretched exponential. Though in the original paper only a hyperbolic kernel was considered, it can be shown that, as in the Dorogovtsev–Mendes model, meaningful results are obtained with nonintegrable kernels. In fact, Fig. 4 shows that the cumulative frequency distribution of tokens displays a fat-tail behavior when kernels are of the type $(1 + \Delta t)^{-\nu}$ with $\nu < 1$. In this model, the Heaps' law is again linear and the power-law frequency distribution of tokens displays an effective exponent not exceeding the value of -2. An overview of the general characteristics and the exponents of the Simon's derived models described above is presented in Table 1.

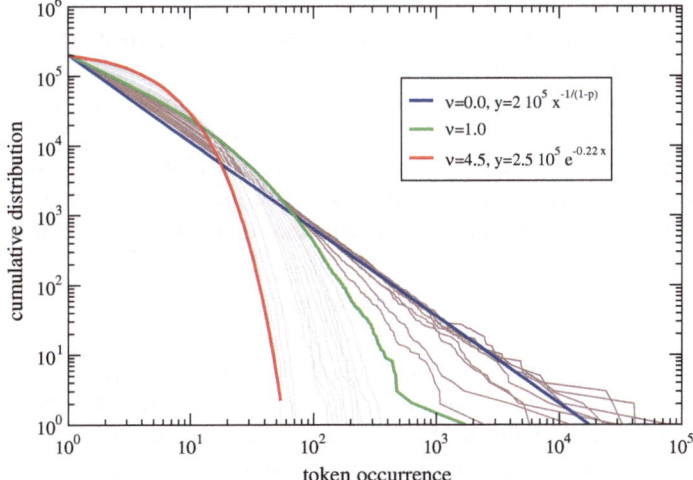

Fig. 4 Behavior of the cumulative distribution of token occurrences in the Cattuto-Loreto-Pietronero (CLP) model [26] as a function of the memory kernel power. A stream with 1,000,000 tokens was generated with a probability of invention $p = 0.2$ and a memory kernel $(1 + \Delta t)^{-\nu}$ at various values of ν. The *blue line* corresponds to the plain Simon's model ($\nu = 0$) where the cumulative frequency distribution of tokens tends asymptotically to the power-law $f^{-1/(1-p)}$; The *brown curves* depict the results for $0 < \nu < 1$; the *green curve* corresponds to the stretched exponential discussed in the CLP paper ($\nu = 1$); the *gray curves* are calculated for $1 < \nu < 4.5$; the *red curve* shows the exponential behavior $\approx e^{-0.22f}$ found with $\nu = 4.5$

Table 1 Comparison of Simon-like models: ZM stands for Zanette–Montemurro [21]; DM for Dorogovtsev–Mendes [25]; CLP for Cattuto-Loreto-Pietronero [26]

Model	Characteristics	Zipf's exponent α	Heaps' exponent
Simon	Constant invention rate p	$1 - p$	1
ZM	Variable invention rate $p = ct^{\gamma-1}$ with $0 < \gamma < 1$	$1/\gamma$	γ
DM	Aging of tokens $\tau^{-\nu}$, constant invention rate	$2 < \alpha < \infty$ for $-\infty < \nu < 1$	1
CLP	Stream aging $(r + \Delta t)^{-1}$, constant invention rate	Stretched exponential with $\alpha \approx 1$	1

3 The Sample-Space Reducing Model

In the attempt to explain fat-tails distributions as observed in real systems, an alternative method to reproduce power-law frequency distributions without explicitly resorting to a rich-gets-richer mechanism was proposed [28]. The model interestingly catches the idea that the space of possibilities often locally reduces when the process goes on: for instance, when composing a sentence the first word is almost free, while the subsequent ones are more and more constrained. The very simple process proposed in [28] works as follows: (i) the process starts with an N-faced dice; (ii) at time t, a j-faced dice resulting from the evolution of the initial N-faced dice is thrown, and let i be the face value obtained; (iii) at time $t + 1$ an i-faced dice is then thrown, and the process goes on until the face value 1 is extracted. It can be shown [28] that, independently of N, the visiting probability for the site i, defined as the probability that a particular process visits the site i before ending at 1, is

$$P_N(i) = \frac{1}{i}. \tag{7}$$

If we consider a cyclic process, i.e., when 1 is reached the process starts again from an N-faced dice, the visiting probability is also proportional to the occupation probability of a site, and thus to the frequency rank, reproducing an exact Zipf law $f(R) \propto R^{-1}$. By relaxing the constraint of deterministically reducing the sample-space, in [28] the authors also study the case in which a probability λ exists to come back at the N-faced dice at each step. In this case, the model can be easily studied as a superposition of the pure sample-space reducing process (with probability $1 - \lambda$) and of a random process in which a number is drawn uniformly in the interval $[1, N]$ (with probability λ). It is shown that one obtains in this case a generalized Zipf law with frequency-rank distribution $f(R) \propto R^{-\lambda}$.

We consider here also the exercise of computing the average number of different values $D(t)$ drawn until the tth step of the cyclic sample-space reducing process described above. This obeys the equation:

$$D(t+1) - D(t) = \sum_i p_i(t) p_{i=\text{new}}(t) \sim \sum_i \frac{A}{i}\left(1 - \frac{A}{i}\right)^t, \tag{8}$$

where $p_i(t)$ is the probability of extracting the number i at step t and $p_{i=\text{new}}(t)$ is the probability that i is extracted at step t for the first time. Note that the last equivalence is not exact since we do not consider the fixed directional constraints imposed by the model. By approximating with the continuous limit, we obtain (see Fig. 5, left panel):

$$\frac{dD(t)}{dt} \sim \sum_i \frac{A}{i}e^{-\frac{tA}{i}} \quad \Rightarrow \quad D(t) = N - \sum_i e^{-\frac{tA}{i}} \tag{9}$$

where A is the normalization constant:

$$A^{-1} = \sum_i^N \frac{1}{i}. \tag{10}$$

Eqs. (8), (9) can be easily generalized when a probability λ to come back to the N-faced dice at each step is introduced:

$$\frac{dD(t)}{dt} = \sum_i (1-\lambda)\frac{A}{i} + \frac{\lambda}{N}\left(1 - \left((1-\lambda)\frac{A}{i} + \frac{\lambda}{N}\right)\right)^t, \tag{11}$$

resulting in (Fig. 5, right panel):

$$D(t) = N - e^{-\frac{\lambda t}{N}} \sum_i e^{-\frac{At}{i}}. \tag{12}$$

Despite its interest, this basic version of the space reducing model is not able to reproduce most of the interesting phenomenology of real processes, in particular a

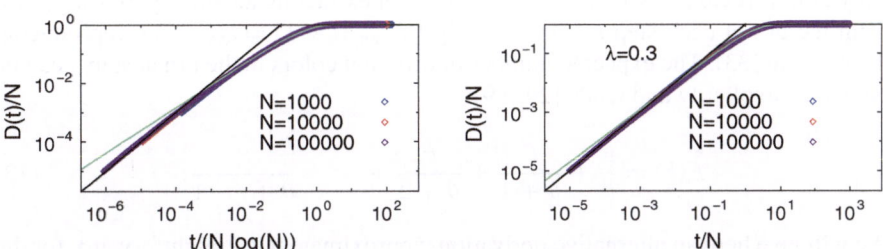

Fig. 5 Heap's law in the sample-space reducing model. *Left* model with $\lambda = 0$. *Right* model with $\lambda = 0.3$. In both cases results are shown for $N = 1000, 10000, 100000$, and the x and y axes are suitably normalized in order to superimpose the *different curves*. The *curves* with both $\lambda = 0$ and $\lambda = 0.3$ are well fitted with a constant plus a stretched exponential function: $f(x) = (1 - \exp(-cx^b))$, with $c = 1.8$, $b = 0.775$ for $\lambda = 0$, and $c = 0.6$, $b = 0.9$ for $\lambda = 0.3$. In both figures the *straight black line* corresponds to $y \propto x$

sublinear power-law behavior for the number of novelties introduced in the process as a function of the number of events, and a Zipf's law with a tail exponent larger than one (in absolute value).

4 Hoppe Urn Model

In 1984, Fred M. Hoppe [29] introduced for the first time the concept of novelties in the framework of urn models [30–32], proposing what is now called the "Hoppe urn model". The inspiration for his work was the Ewens' sampling formula [33]. It describes the allelic partition of a random sampling of n genes from an infinite population at equilibrium, that evolves according to a discrete time neutral Wright-Fisher process [34, 35] with constant mutation rate μ per gene. When taking the infinite population limit $N \to \infty$ and $\mu \to 0$, with $N\mu$ constant, Ewens [33] showed a remarkable property of the sampling process: if one samples j genes from the population at equilibrium, then the probability of sampling the $(j + 1)$th gene with an allele already sampled is $j/(j + \theta)$, where $\theta = 4N\mu$. Hoppe devised his model starting from this property. In the classical version of the Polya urn model [30], balls of various colors are placed in an urn. A ball is withdrawn at random, inspected, and placed back in the urn along with a certain number of new balls of the same color, thereby increasing that color's likelihood of being drawn again in later rounds. The resulting rich-gets-richer dynamics leads to skewed distributions [20, 36] and has been used to model the emergence of power-laws and related heavy-tailed phenomena in fields ranging from genetics and epidemiology to linguistics and computer science [22, 37, 38]. In [29, 39] Hoppe considered a Polya urn with balls of two different qualities: one black ball of mass θ, and colored balls with mass one. The dynamical process works as follows: it starts with only the black ball in the urn, then balls are randomly chosen from the urn proportionally to their mass. At each time step t, (i) if the black ball is extracted, a ball with a brand new color is added to the urn together with the black ball, (ii) if a colored ball is extracted, it is returned to the urn with an additional copy of it. It is easy to see that the probability of extracting an already existing color from the urn at each step $t + 1$ is exactly $P_{\text{existing}}(t + 1) = t/(t + \theta)$, reproducing the result in [33]. The expected number of different colors in the urn at step n can be computed explicitly and reads [29, 39]:

$$E(k) = \frac{\theta}{\theta} + \frac{\theta}{\theta + 1} + \frac{\theta}{\theta + 2} + \cdots + \frac{\theta}{\theta + t - 1}. \tag{13}$$

We will give here an alternative derivation, approximate but straightforward, for the expected number $D(t)$ of different colors in the urn at step t.

The number of different colors $D(t)$ after t extractions follows the recurrence equation:

$$D(t + 1) = D(t) + \frac{\theta}{\theta + t}, \tag{14}$$

where the last term is the probability of extracting a brand new color at time t. We can now take the continuous limit:

$$\frac{dD(t)}{dt} = \frac{\theta}{\theta + t}, \quad D(0) = 0, \tag{15}$$

with solution

$$D(t) = \theta \ln (\theta + t) - \theta \ln (\theta) = \theta \ln \left(1 + \frac{t}{\theta}\right), \tag{16}$$

in accordance with Eq. (13).

In order to compute the frequency-rank distribution $n(R)$, let us call n_i the number of balls with color i in the urn. It follows the equation:

$$n_i(t + 1) = n_i(t) + \frac{n_i(t)}{\theta + t}, \tag{17}$$

where the last term is the probability of extracting the color i at step $t + 1$. Taking again the continuous limit:

$$\frac{dn_i(t)}{dt} = \frac{n_i}{\theta + t}, \quad n(t_i) = 1, \tag{18}$$

where the initial conditions are given at the time t_i when the color i first entered in the urn. By solving the differential equation with the given initial conditions, we obtain the result

$$n_i(t) = \frac{\theta + t}{\theta + t_i} \simeq \frac{t}{\theta + t_i}, \tag{19}$$

where the last approximation holds for $t \gg \theta$. To obtain the probability distribution of the color frequencies, we can now consider

$$P(n_i < n) = P\left(t_i > \frac{t}{n} - \theta\right) = 1 - P\left(t_i < \frac{t}{n} - \theta\right) \simeq 1 - \frac{D(\frac{t}{n} - \theta)}{D(t)}$$

$$= 1 - \frac{\theta \ln \frac{t}{n} - \theta \ln \theta}{\theta \ln t + \theta - \theta \ln \theta} \simeq 1 - \frac{\theta \ln \frac{t}{n} - \theta \ln \theta}{\theta \ln t - \theta \ln \theta} = \frac{\ln n}{\ln t - \ln \theta}, \tag{20}$$

and thus

$$p(n) = \frac{\partial P(n_i < n)}{\partial n} \propto \frac{1}{n}. \tag{21}$$

Finally, the rank R of a color with n occurrences can be obtained by considering

$$R - 1 \simeq k \int_n^{n_{max}} p(n')dn' = kA \ln \frac{n_{max}}{n}, \tag{22}$$

where k is the number of different colors in the urn and A is the normalization constant of the probability distribution $p(n)$:

$$A^{-1} = \int_{n_{min}}^{n_{max}} \frac{1}{n'} dn' = \ln \frac{n_{max}}{n_{min}} = \ln n_{max}, \tag{23}$$

were we set $n_{min} = 1$. By inverting then Eq. (22) we obtain

$$n(R) = n_{max} \exp\left(-\frac{R-1}{k} \ln n_{max}\right), \tag{24}$$

with k given by Eq. (16). In order to estimate n_{max}, let us write the normalization condition

$$\int_1^k n(R') dR' = t. \tag{25}$$

By inserting Eq. (24) in Eq. (25), and noting that

$$n(k) = 1 = n_{max} \exp\left(-\frac{k-1}{k} \ln n_{max}\right), \tag{26}$$

one obtains the relation:

$$k = t \frac{\ln n_{max}}{n_{max} - 1} \simeq t \frac{\ln n_{max}}{n_{max}}, \tag{27}$$

which, by inserting Eq. (16) for k, gives the result $n_{max} \simeq t/\theta$. We thus obtain the estimate:

$$n(R) \simeq \frac{t}{\theta} \exp\left(-\frac{R-1}{\theta}\right). \tag{28}$$

Note that the frequency-rank distribution $n(R)$ was equivalently indicated as $f(R)$ in previous sections.

In Fig. 6 we show results of the Zipf's and Heap's laws, together with their analytic approximations. Again, this model predicts a novelty rate of appearance that is much slower than what is found in many real systems, since the number of different colors $D(t)$ in the urn grows only logarithmically with the number of extractions t.

5   Urn Model with Triggering

In this section, we generalize the urn models seen in the preceding section to propose a modeling scheme that incorporates the notion of the adjacent possible so that one novelty can trigger further novelties. Our approach thus builds on that of Hoppe [29] and other researchers (see Refs. [40, 41] and references therein), who introduced novelties within the framework of Polya's urn but did not posit that they could trigger

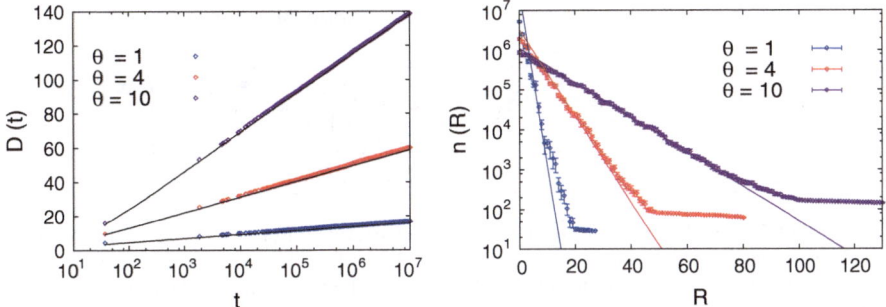

Fig. 6 Zipf's and Heap's laws in the Hoppe urn model. *Left* Heap's law in the Hoppe urn model for different values of the *black* ball mass θ. *Curves* correspond to averages over 100 processes. *Straight lines* correspond to Eq. (16), for each value of θ. *Right* Zipf's law in the Hoppe urn model for different values of the *black* ball mass θ. *Curves* correspond to averages over 100 processes (standard errors are reported). *Straight lines* correspond to Eq. (28), for each value of θ, and for the length of the process $t = 10^7$

subsequent novelties. Hoppe's model was motivated by the biological phenomenon of neutral evolution, with novel alleles represented as an open-ended set of colors arising via mutation from a single fixed color. This variant of Polya's urn implies a logarithmic, rather than power-law, form for the growth of new colors in the urn, and hence does not account for Heaps' law. Hoppe's urn scheme is noncooperative in the sense that no conditional appearance of new colors is taken into account; in particular, one novelty does nothing to facilitate another. In contrast, the cooperative triggering of novelties is essential to our model.

5.1 Model Definition

Consider an urn U initially containing N_0 distinct elements, represented by balls of different colors (Fig. 7). These elements represent songs we have listened to, web pages we have visited, inventions, ideas, or any other human experiences or products of human creativity. A series of inventions is idealized in this framework as a sequence S of elements generated through successive extractions from the urn. Just as the adjacent possible expands when something novel occurs, the contents of the urn itself are assumed to enlarge whenever a novel (never extracted before) element is withdrawn. Mathematically we consider an ordered sequence S, constructed by picking elements (or balls) from a reservoir (or urn), U, initially containing N_0 distinct elements. Both the reservoir and the sequence increase their size according to the following procedure. At each time step:

(i) an element is randomly extracted from U with uniform probability and added to S;
(ii) the extracted element is put back into U together with ρ copies of it;

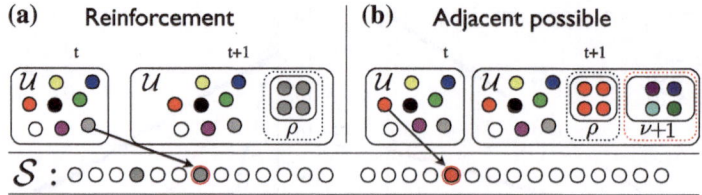

Fig. 7 Models. Simple urn model with triggering. **a** Generic reinforcement step of the evolution. An element (the *gray ball*) that had previously been drawn from the urn U is drawn again. In this case one adds this element to S (depicted at the *center* of the figure) and, at the same time, puts ρ additional *gray balls* into U. **b** Generic adjacent possible step of the evolution. Here, upon drawing a new ball (*red*) from U, $\nu + 1$ brand new balls are added to U along with the ρ *red balls* of the reinforcement step that takes place at each time step

(iii) if the extracted element has never been used before in S (it is a new element in this respect), then $\nu + 1$ different brand new distinct elements are added to U.

Note that the number of elements N of S, i.e., the length $|\mathscr{S}|$ of the sequence, equals the number of times t we repeated the above procedure. If we let D denote the number of distinct elements that appear in S, then the total number of elements in the reservoir after t steps is $|\mathscr{U}|_t = N_0 + (\nu + 1)D + \rho t$.

In parallel with the previous one we consider a slightly different variant of the model, in which the reinforcement does not act when an element is chosen for the first time. Hence, point (ii) of the previous rules will be changed into

(ii.a) the extracted element is put back in U together with ρ copies of it *only if it is not new in the sequence*.

5.2 Computation of the Asymptotic Heaps' and Zipf's Laws

We discuss here the asymptotic behavior of both the number of distinct elements $D(t)$ appearing in the sequence and the frequency-rank distribution $f(R)$ of the elements in the sequence S. We will show that both versions of the urn model above predict a Heaps' law for $D(t)$ and a frequency-rank distribution $f(R)$ with a fat-tail behavior. Our calculations yield simple formulas for the Heaps' law exponent and the exponent of the asymptotic power-law behavior of the frequency-rank distribution in terms of the model parameters ρ and ν.

Strictly speaking, Zipf's law requires an inverse proportionality between the frequency and rank of the considered quantities [42]. In the following, however, we shall always refer instead to a generalized version of Zipf's law, in which the dependence of the frequency on the rank is power-law-like in the tail of the distribution, i.e., at large ranks.

Heaps' law

In the first version of the model, the time dependence of the number D of different elements in the sequence S obeys the following differential equation:

$$\frac{dD}{dt} = \frac{U_D(t)}{U(t)} = \frac{N_0 + \nu D}{N_0 + (\nu + 1)D + \rho t}, \tag{29}$$

where $U_D(t)$ is the number of elements in the reservoir at time t that have not yet appeared in S, and $U(t) = |\mathcal{U}|_t$ is the total number of elements in the reservoir at time t. The term νD in the numerator of the rightmost expression comes from the fact that each time a new element is introduced in the sequence, $U_D(t)$ is increased by ν elements (since $\nu + 1$ brand new elements are added to U, while the chosen element is no longer new). Due to the inherently discrete character of D and t, Eq. (29) is valid asymptotically for large values of D and t.

In the second version of the model, Eq. (29) has to be modified by replacing the denominator with

$$U(t) = N_0 + (\nu + 1)D + \rho(t - D) = N_0 + (\nu + 1 - \rho)D + \rho t.$$

To analyze both versions of the model simultaneously, it is convenient to define a parameter $a \equiv \nu + 1$ for the first version and $a \equiv \nu + 1 - \rho$ for the second version.

In order to obtain an analytically solvable equation, and since we are interested in the behavior at large times $t \gg N_0$, we approximate Eq. (29) by

$$\frac{dD}{dt} = \frac{\nu D}{aD + \rho t}. \tag{30}$$

This equation can be solved by defining $z = \frac{D}{t}$, obtaining:

$$z't + z = \frac{\nu z}{az + \rho}, \tag{31}$$

and thus

$$\int \frac{az + \rho}{z(\nu - \rho - az)} dz = \int \frac{dt}{t}, \tag{32}$$

Solving the integral we obtain

$$\frac{\rho}{\nu - \rho} \log z - \frac{\nu}{\nu - \rho} \log (za + \rho - \nu) = \log t$$

and after some algebra:

$$\frac{z^{\frac{\rho}{\nu}}}{za + \rho - \nu} = t^{\frac{\nu - \rho}{\nu}} \tag{33}$$

By substituting $z = \frac{D}{t}$ and again after some algebra we obtain

$$D^{\frac{\rho}{\nu}} - aD = (\rho - \nu)t, \tag{34}$$

from which we can derive the asymptotic behavior of $D(t)$ for large t:

1. $\rho > \nu$: $D \sim (\rho - \nu)^{\frac{\nu}{\rho}} t^{\frac{\nu}{\rho}}$;

2. $\rho < \nu$: $D \sim \frac{\nu - \rho}{a}t$;

3. $\rho = \nu$: $D \log D \sim \frac{\nu}{a}t \rightarrow D \sim \frac{\nu}{a}\frac{t}{\log t}$,

where the last estimate cannot be deduced directly from Eq. (34), but is deduced by substituting $\nu = \rho$ directly in Eq. (31).

In the case $\rho < \nu$, we recover the results of the well-known Simon's Model [20], originally proposed in the context of linguistics and described in Sect. 2. The Simon's model leads to a Zipf's law with an exponent $-(1 - p)$ compatible with a linear growth in time of the number of different words. In the framework of the present *urn model with triggering* we recover the same Zipf's exponents as well as the linear growth of $D(t)$ if $p = 1 - \frac{\rho}{\nu}$, with $\rho < \nu$.[1]

For completeness, we note that both versions of the model can be regarded as the coarse-grained equivalent of a two-color asymmetric Polya urn model [32]. In particular, within that finer framework the substitution matrices (denoted M_1 for the first version of the model and M_2 for the second) would be

$$M_1 = \begin{pmatrix} \rho & 0 \\ 1 + \rho & \nu \end{pmatrix} \quad \text{and} \quad M_2 = \begin{pmatrix} \rho & 0 \\ 1 & \nu \end{pmatrix}.$$

In this interpretation, the elements that have already appeared in S are represented by balls of one color, while those that have not appeared yet correspond to balls of the other color.

Zipf's law

Making the same approximations as above, the continuous dynamical equation for the number of occurrences n_i of an element i in the sequence S can be written as

$$\frac{dn_i}{dt} = \frac{n_i \rho + 1}{N_0 + aD + \rho t}. \tag{35}$$

Two cases can be distinguished:

---

[1] We note that if $\nu \gg 1$ when $a = \nu + 1$ (first version of the model) or $\nu \gg \rho$ and $\nu \gg 1$ when $a = \nu + 1 - \rho$ (second version of the model) our model also reproduces the same prefactor of the linear growth of $D(t)$ as in Simon's model. This is evident by setting $a = \nu$ in Eq. (30).

1. $v \leq \rho$, when $\lim_{t \to +\infty} D/t = 0$. By considering only the leading term for $t \to +\infty$, one has

$$\frac{dn_i}{dt} \simeq \frac{n_i}{t}. \tag{36}$$

Let t_i denote the time at which the element i occurred for the first time in the sequence. Then the solution for $n_i(t)$ starting from the initial condition $n_i(t_i) = 1$ is given by

$$n_i = \frac{t}{t_i}. \tag{37}$$

Now consider the cumulative distribution $P(n_i \leq n)$. From Eq. (37), we can write $P(n_i \leq n) = P(t_i \geq \frac{t}{n}) = 1 - P(t_i < \frac{t}{n})$. This leads to the estimate:

$$P\left(t_i < \frac{t}{n}\right) \simeq \frac{D(\frac{t}{n})}{D(t)} = n^{-\frac{v}{\rho}}. \tag{38}$$

2. $v > \rho$, when $D \simeq \frac{v-\rho}{a} t$. Again considering $t \gg N_0$, we write

$$\frac{dn_i}{dt} \simeq \frac{\rho n_i}{(\rho + a\frac{v-\rho}{a})t} = \frac{\rho n_i}{vt}, \tag{39}$$

which yields the solution

$$n_i = \left(\frac{t}{t_i}\right)^{\frac{\rho}{v}}. \tag{40}$$

Proceeding as in the previous case, we find $P(n_i \leq n) = P(t_i \geq t n^{-\frac{v}{\rho}}) = 1 - P(t_i < t n^{-\frac{v}{\rho}})$, and thus

$$P(t_i < t n^{-\frac{v}{\rho}}) \simeq \frac{D(t n^{-\frac{v}{\rho}})}{D(t)} = n^{-\frac{v}{\rho}}, \tag{41}$$

obtaining the same functional expression of the asymptotic power-law behavior of the frequency-rank distribution as in the previous case.

The probability density function of the occurrences of the elements in the sequence is therefore $P(n) = \frac{\partial P(n_i \leq n)}{\partial n} \sim n^{-\left(1+\frac{v}{\rho}\right)}$, which corresponds to a frequency-rank distribution $f(R) \sim R^{-\frac{\rho}{v}}$.

Figure 8 shows the theoretical predictions and the numerical simulations for the Heaps' and Zipf laws. The robustness of the results with respect to fluctuations of the model parameters v and ρ was also checked by sampling their values out of several probability distributions. Note that the estimates in Eqs. (38) and (41) have been derived under the assumption that $t/n \gg 1$, i.e., in the tail of the frequency-rank distribution. In this respect, it is important to recognize that Zipf's and Heaps' laws

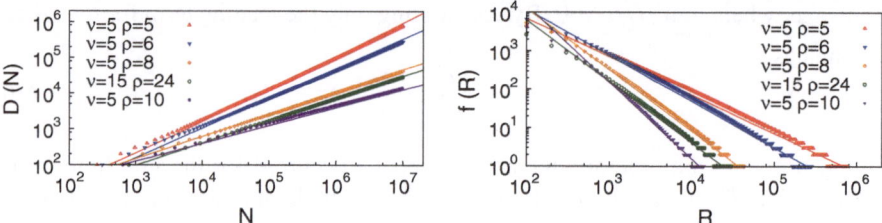

Fig. 8 Heaps' law (*left*) and Zipf's law (*right*) in the urn model with triggering. *Straight lines* in the Heaps' law plots show functions of the form $f(x) = ax^\gamma$ with the exponent $\gamma = \nu/\rho$ as predicted by the analytic results and confirmed in the numerical simulations. *Straight lines* in the Zipf's law plots show functions of the form $f(x) = ax^{-\alpha}$, where the exponent α is equal to γ^{-1}. Note that the frequency-rank plots in real data deviate from a pure power-law behavior and the correspondence between the γ and α exponents is valid only asymptotically

are not trivially and automatically related, as is sometimes claimed. We certainly agree that Heaps' law can be derived from Zipf's law by the following random sampling argument: if one assumes a strict power-law behavior of the frequency-rank distribution $f(R) \sim R^{-\alpha}$ and constructs a sequence by randomly sampling from this Zipf distribution $f(R)$, one recovers Heaps' law with the functional form $D(t) \sim t^\gamma$ with $\gamma = 1/\alpha$ [43, 44]. But the assumption of random sampling is strong and sometimes unrealistic. If one relaxes the hypothesis of random sampling from a power-law distribution, the relationship between Zipf's and Heaps' law becomes far from trivial. In our model, and in work by others [44], the relationship $\gamma = 1/\alpha$ holds only asymptotically, i.e., only for large times, with α measured on the tail of the frequency-rank distribution.

6 Conclusions

The processes leading to the emergence of novelties and innovations are mostly unknown. Still, the observation of statistical regularities displayed by the occurrence of new events is a key to chart the unknown territories that describe the space of possibilities for societies, biological systems, and technology. In this short and far from exhaustive review, we made an attempt to draw a path through the attempts recently made to model, with tools borrowed from the theory of complex systems, the emergence of novelties and innovations. To this end we considered a strip of the most renowned models, proposed in almost a century of studies. Some models were historically treated without the explicit aim of modeling the occurrence of novelties and innovations. For example, the model of Simon (see Sect. 2) was conceived to reproduce the frequency distribution words display in texts and in fact in Simon's model the rate of novelty creation is constant, while in many real systems, including texts, it decreases in time with a power-law behavior. Nevertheless, in light of the results of the new model of Polya's urn with triggering (Sect. 5), Simon's model turns out to be correctly describing a case in which the space of possibilities grows at a fast pace ($\nu > \rho$). By ad-hoc inserting a sublinear rate of inventions in Simon's model, Zanette and Montemurro [21] obtained a satisfactory description of both the frequency distribution of words in texts and the rate of occurrence of new words. Of course the latter is a trivial consequence of having imposed the correct rate *deus ex machina*. Simon's model can be considered a milestone for all models based on the construction of a stream of tokens (e.g., those models involving memory effects, nonlinear preferential attachment, etc.) and its limited ability in reproducing real data has been perfected with the idea of using Polya's urns. Polya's urns are well known to mathematicians who developed a multitude of techniques to cope with them, for instance looking at urns containing balls with a finite number of colors, i.e., with a finite space, and nontrivial transition probabilities. One step forward is that of introducing a simple way to enlarge the space of possibilities. To this end, the Hoppe–Polya model (Sect. 4) already represents a good solution, though the rate of occurrence of innovations is still too low and far from the actually observed values in many systems of interest. Finally, the model of Polya's urn with innovation triggering (Sect. 5), formalizing the notion of adjacent possible envisioned by S. Kauffman, presents for the first time a satisfactory first principle-based way of reproducing empirical observations. Not only both Heaps' and Zipf's laws of real case situations are reproduced, but also the classical model of Simon is retrieved in the limit of fast growth of the space.

In a somewhat humorously self-referring sense, each proposed model has been in the adjacent possible of the models prior to it. But of course this is only an a posteriori consideration. Nobody knows what the adjacent possible space looks like and even conceptually it is not clear which tools one could possibly adopt to chart it. From this perspective, we hope that the recent stream of investigations connected to the Polya's urn presented in Sect. 5, by providing the first quantitative characterization of the dynamics of correlated novelties, could be a starting point for a deeper understanding of the different nature of triggering events (timeliness, scales, spreading, individual versus collective properties) along with the signatures of the adjacent possible at the individual and collective level, its structure and its restructuring under individual innovative events.

Appendix

Relation Between Frequency Distribution and Frequency-Rank Distribution (Zipf's Law)

Let us consider a sequence of N random variables (or letters, or any kind of item), and the frequency at which each particular value enters in the sequence. For many systems (genes in a pan-genome, words in a texts, etc.), the distribution of the number of appearances of the same value of the variable in a sequence is a power-law:

$$p(f) = (\beta - 1)f_{\min}^{\beta-1}f^{-\beta}, \tag{42}$$

with $\beta > 1$.

Let us now relate the *frequency distribution* to the *frequency-rank distribution*: the elements in the sequence are ordered according to their frequency in decreasing order (rank one for the most frequent element), and the frequency is studied as a function of the rank R of each element.

Let us now compute the form of frequency-rank curve (also called Zipf's law) corresponding to the frequency distribution in Eq. (42). The rank R of an element with frequency f is defined as the number of different elements with frequency $\tilde{f} \geq f$:

$$R(f) \simeq k \int_f^{+\infty} p(\tilde{f})d\tilde{f}, \tag{43}$$

where k is the number of distinct elements in the sequence. By substituting Eq. (42) into Eq. (43) we obtain:

$$R(f) \propto f^{1-\beta} \tag{44}$$

and inverting the relation:

$$f(R) \propto R^{-\alpha} \qquad \text{with} \qquad \alpha = \frac{1}{\beta - 1}. \tag{45}$$

In order to obtain the correct expression for $f(R)$, we now have to consider the normalization:

$$\int_1^{R_{\max}} f(\tilde{R})d\tilde{R} = 1. \tag{46}$$

Let us now distinguish the cases:

$\alpha \neq 1 \ (\beta \neq 2)$:

$$f(R) = \frac{1-\alpha}{R_{\max}^{1-\alpha} - 1}R^{-\alpha}. \tag{47}$$

$\alpha = 1 \ (\beta = 2)$:

$$f(r) = \frac{n}{\ln r_{\max}}r^{-1}. \tag{48}$$

Let us now note that when $\alpha > 1$, we can neglect the term $R_{\max}^{1-\alpha}$ in Eq. (47), and when $\alpha < 1$, we can write $R_{\max}^{1-\alpha} - 1 \simeq R_{\max}^{1-\alpha}$.

Summarizing:

$$\alpha > 1 \ (\beta < 2): \quad f(R) \simeq (\alpha - 1)R^{-\alpha}. \tag{49}$$

$$\alpha = 1 \ (\beta = 2): \quad f(R) \simeq \frac{R^{-1}}{\ln R_{\max}}. \tag{50}$$

$$\alpha < 1 \ (\beta > 2): \quad f(R) \simeq (1 - \alpha)\frac{R^{-\alpha}}{R_{\max}^{1-\alpha}}. \tag{51}$$

Relation Between Frequency-Rank Distribution and Number of Distinct Elements (Heaps Law)

We now want to estimate the number D of distinct elements appearing in the sequence as a function of its length N (Heaps' law). To do that, let us consider the entrance of a new element (never appeared before) in the sequence and let the number of distinct elements in the sequence be D after this entrance, and the length of the sequence N. This new element will have maximum rank $R_{\max} = D$, and frequency $f(R_{\max}) = 1/N$. From Eqs. (49)–(51) we thus obtain:

$$\alpha > 1 \ (\beta < 2): \quad f(D) \simeq (\alpha - 1)D^{-\alpha} = \frac{1}{N}. \tag{52}$$

$$\alpha = 1 \ (\beta = 2): \quad f(D) \simeq \frac{1}{D \ln D} = \frac{1}{N}. \tag{53}$$

$$\alpha < 1 \ (\beta > 2): \quad f(D) \simeq \frac{1 - \alpha}{D^{1-\alpha} - 1}D^{-\alpha} = \frac{1}{N}. \tag{54}$$

Inverting these relations we finally find:

$$\alpha > 1 \ (\beta < 2): \quad D \simeq N^{\gamma} \quad \text{with} \quad \gamma = \frac{1}{\alpha}. \tag{55}$$

$$\alpha = 1 \ (\beta = 2): \quad D \simeq \frac{N}{\ln N}. \tag{56}$$

$$\alpha < 1 \ (\beta > 2): \quad D \simeq N. \tag{57}$$

Some Remarks About Text Generation

The production of written text in a given idiom is a peculiar process. Although texts are generally taken as a paradigmatic example of a collection of tokens (words) that obey both Zipf's and Heaps' laws and many modeling efforts have been undertaken to explain this feature, writers very seldom create brand new words in that idiom. Rather, the process of text production is well approximated by the sampling of the existing Zipf's law of that idiom at the specific time of its writing. Authors, then, choose words according to their frequency and add the necessary short range correlations needed to compose meaningful sentences and long range correlations to follow the plot line.

The generation of the Zipf's law of words in a given idiom is therefore a different process with respect to book writing and the scientific data analysis and modeling efforts have to be steered toward the comprehension of this global shared law. Notably, some work has been already done based on *Google ngrams* and it was shown that the Zipf's distribution of English varies from decade to decade [45] but in a way that its functional form stays constant. The functional form is that of a double slope in logarithmic scale (see Fig. 2d for the Zipf's law inferred from the Gutenberg corpus) with the Heaps' law connected to the tail of the distribution, i.e., to the appearance of less frequent words, as it should be. A satisfactory model accounting for the frequency-rank distribution of words in a language has to include different ingredients, therefore, from the mechanism of creation of neologisms, to the aging and disappearance of terms and revival of others. Above all, the mechanism has to highlight the collaborative and shared character of idioms.

References

1. North, M.: Novelty: A History of the New (University of Chicago Press, Chicago 2013)
2. Zabell, S.L.: Synthese **90**(2), 205 (1992)
3. Jacob, F.: The Possible and the Actual. University of Washington Press, Washington (1982)
4. Kauffman, S.A.: The Origins of Order: Self-Organization and Selection in Evolution. Oxford University Press, New York (1993)
5. Kauffman, S.A.: Investigations. Oxford University Press, New York (2000)
6. Johnson, S.: Where Good Ideas Come From: The Natural History of Innovation. Riverhead Hardcover, New York (2010)
7. Kauffman, S., Thurner, S., Hanel, R.: Scientific American (online) (2008)
8. Thurner, S., Klimek, P., Hanel, R.: New J. Phys. **12**, 075029 (2010)
9. Solé, R.V., Valverde, S., Casals, M.R., Kauffman, S.A., Farmer, D., Eldredge, N.: Complexity **18**(4), 15 (2013)
10. Felin, T., Kauffman, S.A., Koppl, R., Longo, G.: Strateg. Entrep. J. **8**(4), 269 (2014)
11. Buchanan, M.: Nat. Phys. **10**, 243 (2014)
12. Tria, F., Loreto, V., Servedio, V.D.P., Strogatz, S.H.: Nature Scientific Reports **4** (2014)
13. Hart, M.: Project Gutenberg (1971). http://www.gutenberg.org/
14. Schachter, J.: del.icio.us (2003). http://delicious.com/
15. De Morgan, A.: An Essay on Probabilities, and Their Application to Life Contingencies and Insurance Offices. Longman et al, London (1838)
16. Estoup, J.B.: Institut Stenographique de France (1916)
17. Condon, E.U.: Science **67**, 300 (1928)
18. Zipf, G.K.: The Psychobiology of Language. Houghton-Mifflin, New York (1935)
19. Ferrer-i Cancho, R., Elvevag, B.: PLoS ONE **5**(3), e9411 (2010)
20. Simon, H.A.: Biometrika **42**, 425 (1955); B. Mandelbrot, Information and Control **2**, 90 (1959); H.A. Simon, Information and Control **3**, 80 (1960); B. Mandelbrot, Information and Control **4**, 198 (1961); H.A. Simon, Information and Control **4**, 217 (1961); B. Mandelbrot, Information and Control **4**, 300 (1961); H.A. Simon. Information and Control **4**, 305 (1961)

21. Zanette, D., Montemurro, M.: J. Quant. Ling. **12**, 29 (2005)
22. Simkin, M.V., Roychowdhury, V.P.: Phys. Rep. **502**, 1 (2011)
23. Heaps, H.S.: Information Retrieval-Computational and Theoretical Aspects (Academic Press, Cambridge 1978)
24. Barabási, A.L., Albert, R.: Science **286**, 509 (1999)
25. Dorogovtsev, S.N., Mendes, J.F.F.: Phys. Rev. E **62**, 1842 (2000)
26. Cattuto, C., Loreto, V., Pietronero, L.: Proc. Natl. Acad. Sci. **104**, 1461 (2007)
27. Cattuto, C., Loreto, V., Servedio, V.D.P.: Eur. Phys. Lett. **76**, 208 (2006)
28. Corominas-Murtra, B., Hanel, R., Thurner, S.: Proc. Natl. Acad. Sci. **112**(17), 5348 (2015)
29. Hoppe, F.M.: J. Math. Biol. **20**(1), 91 (1984)
30. Pólya, G.: Annales de l'I.H.P. **1**(2), 117 (1930)
31. Johnson, N.L., Kotz, S.: Urn Models and Their Application: An Approach to Modern Discrete Probability Theory (Wiley, New York 1977)
32. Mahmoud, H.: Pólya Urn Models. Texts in Statistical Science Series. Taylor and Francis Ltd, Hoboken (2008)
33. Ewens, W.: Theor. Popul. Biol. **3**, 87 (1972)
34. Fisher, R.A.: The Genetical Theory of Natural Selection. Clarendon Press, Oxford (1930)
35. Wright, S.: Genetics **16**(2), 97 (1931)
36. Yule, U.G.: Philosophical Transactions of the Royal Society of London. Series B, Containing Papers of a Biological Character **213**, 21 (1925)
37. Mitzenmacher, M.: Internet Math. **1**, 226 (2003)
38. Newman, M.E.J.: Contemp. Phys. **46**, 323 (2005)
39. Hoppe, F.M.: J, Math. Biol. **25**(2), 123 (1987)
40. Kotz, S., Balakrishnan, N.: Advances in combinatorial methods and applications to probability and statistics. In: Balakrishnan, N. (ed.) Statistics for Industry and Technology, pp. 203–257. Birkäuser, Boston (1996)
41. Alexander, J.M., Skyrms, B., Zabell, S.: Dyn. Games Appl. **2**(1), 129 (2012)
42. Zipf, G.K.: Human Behavior and the Principle of Least Effort. Addison-Wesley, Reading (1949)
43. Serrano, M.A., Flammini, A., Menczer, F.: PLoS ONE **4**(4), e5372 (2009)
44. Lü, L., Zhang, Z.K., Zhou, T.: PLoS ONE **5**(12), e14139 (2010)
45. Gerlach, M., Altmann, E.G.: Phys. Rev. X **3**, 021006 (2013)

Generating Non-plagiaristic Markov Sequences with *Max Order* Sampling

Alexandre Papadopoulos, François Pachet and Pierre Roy

Abstract Plagiarism is usually studied from an analysis viewpoint: how to detect that a text contains copies of another one. In this chapter we study plagiarism from the generation viewpoint: how to generate a text with a guarantee of non-plagiarism. More precisely, we address the problem of Markov sequence generation with forbidden k-gram constraints. This problem is addressed in two steps. In the first step, we show that, given a Markov transition matrix and a set of k-grams, we can build efficiently an automaton that represents exactly the language of all sequences that can be generated from a Markov model, and that also do not contain any of the k-grams. The size of the automaton is bounded by the size of the forbidden k-grams, and so is the time for building it. This automaton can be used to solve the algebraic problem (i.e. considering non-zero probabilities are uniform), by a simple walk. In the second step, we show that the automaton can be extended so as to be exploited by a belief propagation scheme, in order to produce perfect sampling of all the solutions.

1 Introduction

Markov chains are a powerful, widely-used technique to analyse and generate sequences that imitate a given style [4, 13], with applications to many areas of automatic content generation such as music, text and more generally sequential data. A typical use of such models is to generate *novel* sequences that "look" like or "sound" like the original.

From a corpus of finite-length sequences considered as representative of the style of an author, a Markov model of the style is estimated based on the Markov

A. Papadopoulos (✉)
UPMC Paris 6, UMR 7606, LIP6, 75005 Paris, France
e-mail: alexandre.papadopoulos@lip6.fr

F. Pachet · P. Roy
Sony CSL, 6 rue Amyot, 75005 Paris, France
e-mail: pachetcsl@gmail.com

P. Roy
e-mail: roypie@gmail.com

© Springer International Publishing Switzerland 2016   85
M. Degli Esposti et al. (eds.), *Creativity and Universality in Language*,
Lecture Notes in Morphogenesis, DOI 10.1007/978-3-319-24403-7_6

hypothesis, which states that the future state of a sequence depends only on the last state, i.e.:

$$P(s_{i+1}|s_1, \ldots, s_i) = P(s_{i+1}|s_i).$$

The equation above describes a Markov model of order 1. The definition can be extended to higher orders by considering prefixes of length k greater than 1.

$$P(s_{i+1}|s_1, \ldots, s_i) = P(s_{i+1}|s_{i-k+1}, \ldots, s_i).$$

Variable order Markov models are often used to produce sequences with varying degrees of similarity with the corpus [2]: an increased Markov order produces sequences that replicate larger chunks of the original corpus, thereby improving the impression of style imitation. However, a Markov process can yield long-range effects, in spite of its local nature. The result of such effects is that undesirable substrings of a length greater than the Markov order can be generated, with a high frequency. For example, it has been long observed [4] that increasing the order tends to produce sequences that contain chunks of the corpus of size much larger than the Markov order.

We illustrate this phenomenon on a text corpus: Johnston's English translation of Pushkin's Eugene Onegin—a reference to Markov, as he used the same corpus (in Russian) for his pioneering studies. Here, an element of the Markov chain is a word of the text or a sentence separator, and a sequence is a succession of such elements. We obtained the following sequence with a Markov order of 1:

Praskovya re-baptized "Polina". Walking her secret tome that rogue, backbiter, pantaloon, bribe-taker, glutton and still eats, and featherbeds, and enjoyment locked him all went inside a day wood below the flower was passion and theirs was one who taught her handkerchief has measured off in caravan the finest printer with pulsesracing down, he'll be nothing could draw it abounded.

On top of the text, we draw the longest subsequences that appear verbatim the corpus, or *chunks*, assigning different colours to different lengths. For example, this generated sequence contains the chunk "[...] that rogue, backbiter, pantaloon, bribe-taker, glutton and [...]", which is a seven-word-long subsequence from the corpus. The maximum order of a sequence is the maximum length of its chunks (seven, in our example).

With an increased Markov order of 3, we obtained the following sequence:

Love's frantic torments went on beating and racking with their strain and stress that youthful heart. It all seemed new – for two days only – the estate provides a setting for angry heirs, as one, to admire him – and replies: "Wait, I'll present you – but inside a day, with custom, love would fade away". It's right and proper that you transcend in music's own bewitching fashion the foreign words a maiden's passion found for its utterance that night directed his.

This sequence makes, locally, more sense than the one generated with order 1. However, its maximum order is 20 (i.e. it contains a twenty-word-long subsequence copied verbatim from the corpus). To any reader familiar with the corpus, this would read like blatant plagiarism.

More generally, we generated a few hundreds of sequences of size 30 based on this corpus, of varying order (from to 1 to 3): Fig. 1 shows the distribution of chunk sizes observed for each Markov order. With a Markov order 2, sequences already tend to contain chunks from the corpus of length greater than 2, up to 22. Markov order affects *training*: when estimated from a corpus, the Markov model learns all continuations of sequences of k states or less. However, this parameter k does not limit the maximum order of the generated sequence.

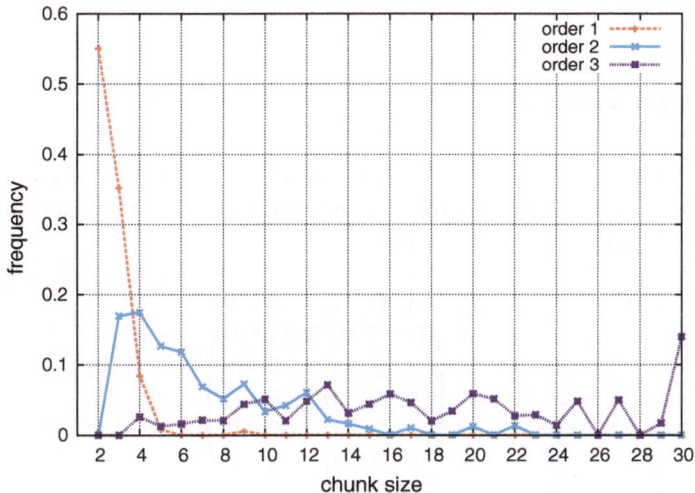

Fig. 1 Chunk size distribution for different Markov orders in a sequence of size 30

To avoid this type of plagiarism, we are interested in generating Markov sequences with a guaranteed maximum order. In other words, we want to forbid all sequences of length equal to the maximum order, that appear in the corpus. But other types of undesirables sequences can occur, too. For example, anti-patterns [5] are minimal sequences (in terms of size) that do not appear in the corpus, and yet have a high probability according to the model estimated on the corpus. Although a Markov process will reproduce anti-patterns with a high probability, it can be argued that their absence in the corpus has a structural justification, and, therefore, we may want to explicitly prevent their occurrences in generated sequences. Cyclical patterns are small sequences of words repeated several times successively. Although long cyclical patterns are unlikely, short cyclical patterns have a non-trivial probability of occurring, and make no sense when generating text. If a cyclical pattern has a period longer than the Markov order, again we need another means to explicitly forbid it.

This chapter addresses precisely the problem of sampling Markov sequences that contain no forbidden substring, or *no-good* for short. Such properties cannot be guaranteed with greedy approaches like random walk. Our contribution combines techniques from constraint satisfaction, automaton theory and statistical inference. From a satisfaction viewpoint, we consider a constraint enforcing that generated sequences contain none of the imposed no-goods. Following a common approach in constraint programming [3, 12], we represent such a constraint with an automaton that accepts the set of such sequences. However, we show that canonical methods are not satisfactory for building this automaton. We present an algorithm that builds this automaton in linear time with respect to the set of no-goods, an aspect that has been covered in detail in previous work [9]. In a second, novel, step, this automaton is used to define a factor graph that encodes the distribution of Markov sequences containing no no-good. Belief propagation is then applied to sample such sequences in an exact manner.

2   A Simple Example

We consider the corpus made of ABRACADABRA, where each element is one of the symbols A, B, C, D or R. With $k = 1$, the Markov chain estimated on this corpus is given by the following transition matrix:

$$
\begin{array}{c c}
 & \begin{array}{c c c c c} A & B & C & D & R \end{array} \\
\begin{array}{c} A \\ B \\ C \\ D \\ R \end{array} &
\left(\begin{array}{c c c c c}
0 & 0.5 & 0.25 & 0.25 & 0 \\
0 & 0 & 0 & 0 & 1 \\
1 & 0 & 0 & 0 & 0 \\
1 & 0 & 0 & 0 & 0 \\
1 & 0 & 0 & 0 & 0
\end{array}\right)
\end{array}
$$

During a training phase, these probabilities are estimated according to their frequency in the corpus. Here, in the four continuations for A in the corpus, two are with B, one with C and one with D. A sequence is a Markov sequence, according to an order k Markov chain, if every k-gram of the sequence has a continuation with a non-zero probability. For example, ABRADABRACA is a valid Markov sequence, but ABRACADABA is not a valid Markov sequence, because the probability of having A after B is zero.

We can encode a set of Markovian sequences, ignoring probabilities, using an *automaton*.

Definition 1 (*Automata*) A deterministic finite-state automaton, or, simply, automaton, is a quintuple $\mathcal{A} = \langle Q, \Sigma, \delta, q_0, F \rangle$, where Q is a finite non-empty set of states, Σ, the alphabet, is a finite non-empty set of symbols, $q_0 \in Q$ is the initial state, δ is the transition function which maps a state q and a symbol a to the successor $\delta(q, a)$, and $F \subseteq Q$ is the set of final, or accepting, states.

Definition 2 (*Accepted language*) The word w, a string of symbols in Σ, is accepted by $\mathcal{A}$ if and only if there exists a sequences $q_0, \ldots, q_p$ of states, such that $\delta(q_{i-1}, a_i) = q_i$, for all i, $1 \leq i \leq p$, and $q_p \in F$. The set of words accepted by $\mathcal{A}$, denoted $\mathcal{L}(\mathcal{A})$, is the language accepted by $\mathcal{A}$.

In order to represent an order k Markov chain, we create an alphabet Σ where each symbol corresponds to a state of the Markov chain, a k-gram. Then, a valid order k Markov transition is represented by two symbols, such that their two corresponding k-grams overlap on their common $k - 1$ symbols. A valid Markov sequence of length n is represented by a word of length $n - k + 1$ on this alphabet. For example, for $k = 2$, the sequence ABRA corresponds to a sequence of three 2-grams $\langle A, B \rangle$, $\langle B, R \rangle$, $\langle R, A \rangle$. We can assign three symbols $a_1, a_2, a_3 \in \Sigma$ to those three 2-grams in their respective order. The Markov transition $A, B \rightarrow R$ is represented by the word a_1a_2, and the sequence ABRA by the word $a_1a_2a_3$.

Definition 3 (*Markov Automaton*) A Markov automaton associated to a Markov chain is an automaton $\mathcal{A}$, such that in each accepted word $a_1 \ldots a_p$ in $\mathcal{L}(\mathcal{A})$, for each $i < p$, a_ia_{i+1} corresponds to a non-zero probability Markov transition.

Figure 2 shows a Markov automaton for the Markov chain estimated on ABRACADABRA with $k = 1$. This automaton is essentially a syntactic rewrite of the Markov chain, with a different semantics attached to its states and transitions. Since the automaton accepts sequences of arbitrary length, all states are accepting. Each transition is labelled by a symbol corresponding to a Markov state. A notable property of a Markov automaton is that all transitions labelled with a given symbol point to the same state. Furthermore, we can impose, at the expense of minimality, that all transitions pointing to a given state are labelled with the same symbol. For example, on Fig. 2, all transitions labelled with R point to the same state, but transitions labelled with D and C also point to this state. In order to enforce the second invariant too, we need to make three copies of this state, for C, D and R. Satisfying those two invariants allows us to introduce the following notation.

Fig. 2 A Markov automaton
for the ABRACADABRA
corpus, with $k = 1$

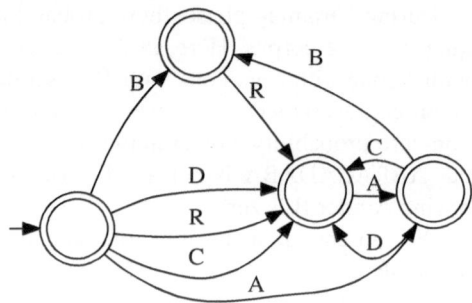

Definition 4 We relate states and the labels of its ingoing transitions, using the following notation:

- Let a be a symbol of the alphabet, $Q(a)$ is the *unique* state q to which a-labelled transitions point.
- Let q be a state in Q, $a(q)$ is the label of the transitions pointing to q, i.e. $Q(a(q)) = q$.

Using this notation, a Markov transition between the k-grams corresponding to a_1 and a_2 is represented in $\mathcal{A}$ by a transition between $Q(a_1)$ and $Q(a_2)$, labelled by a_2. We can now represent the set of Markov sequences containing no forbidden no-good with the following automaton.

Definition 5 (*No-good Automaton*) Let $\mathcal{M}$ be a Markov automaton, and $\mathcal{N}$ a list of strings accepted by $\mathcal{M}$, called *no-goods*.

For a given no-good $a_1 \ldots a_L \in \mathcal{N}$, let $\mathcal{A}(a_1 \ldots a_L)$ be an automaton such that $\mathcal{L}(\mathcal{A}(a_1 \ldots a_L)) = \{w \in \Sigma^* \mid a_1 \ldots a_L \text{ is a substring of } w\}$, i.e. the language of words containing at least one occurrence of the no-good. An automaton $\mathcal{NG}$ is a no-good automaton for $\mathcal{M}$ and $\mathcal{N}$ if:

$$\mathcal{L}(\mathcal{NG}) = \mathcal{L}(\mathcal{M}) \cap \bigcap_{a_1 \ldots a_L \in \mathcal{N}} \overline{\mathcal{L}(\mathcal{A}(a_1 \ldots a_L))}$$

For example, in the ABRACADABRA corpus, with $k = 1$ and a max order limit of 4, we have 7 max order no-goods: ABRA, BRAC, RACA, ACAD, CADA, ADAB, DABR. The Markovian sequence ABRADABRACA does not satisfy the maximum order property: it contains the no-goods ABRA, ADAB, DABR, BRAC, RACA. The Markovian sequence RADADACAB does not contain any no-good of size 4, and so satisfies the maximum order 4 property: its max order is strictly less than 4. Figure 3 shows a no-good automaton, forbidding those no-goods. Again, all states are accepting, and the labels in the states correspond to prefixes of forbidden

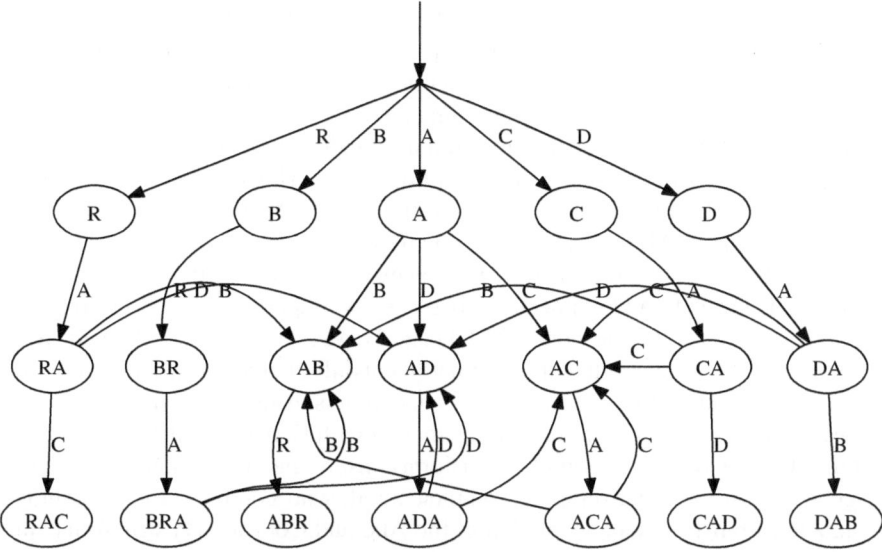

Fig. 3 A no-good automaton for the ABRACADABRA corpus, with $k = 1$ and maximum order less than 4

no-goods. Note that all transitions pointing to any given state are labelled with the same symbol.

3 Building the No-Good Automaton

The no-good automaton can be built in a generic fashion, using standard automata theory operations that implement Definition 5. Initially, we build a Markov automaton. Then, for each no-good, this automaton is intersected with the negation of the automaton recognising sequences containing the no-good. The complexity of this procedure is dominated by the complexity of intersecting a number of automata. A straightforward intersection algorithm runs in $O(t^N)$ time, with t the maximum size, in number of states, of any of the automata, and $N = |\mathcal{N}|$ the number of no-goods. And it is unlikely that an algorithm with a better complexity exists [7, 8]. Therefore, we cannot tractably compute the no-good automaton in a generic way, without exploiting its particular structure. Furthermore, this method does not give any bound on the size of the final automaton (other than $O(t^N)$).

We present, in this section, an algorithm that builds the no-good automaton in time linear in the size of the set of no-goods. As a corollary, we show that the size of this automaton is linear too. More precisely, this algorithm takes as input a Markov automaton and a list of no-goods, and operates as follows: build a trie with the no-goods, compute their overlaps, and finally use this structure to remove from the Markov automaton all sequences containing a no-good.

Algorithm 1 first computes a trie of the no-goods, where all states but the ones corresponding to a full no-good are accepting states. This ensures that a no-good is never accepted. This trie is connected to the original Markov automaton by disconnecting from the Markov automaton any transition that starts a no-good, and use those transitions to start building the trie. However, this is not sufficient. The key part of the algorithm is to add connections between overlapping no-good prefixes. For example, if we have two no-goods ABCD and BCEF, the prefixes ABC and BCEF overlap on BC. This means that the automaton not only should not accept ABCD, but it should not accept ABCEF either. This connection is made using *cross-prefix transitions*: we add a cross-prefix transition, labelled with E, between the state for ABC and the state for BCE. Cross-prefix transitions ensure that, by avoiding a particular no-good, we will not inadvertently complete another no-good. In order to compute them, Algorithm 1 uses an adaptation of the Aho and Corasick [1] string-matching algorithm: a transition that does not extend the no-good of the current no-good prefix is a cross-prefix transition, and points directly to the no-good that starts with the longest suffix of the current prefix. Finally, we add transitions in the trie for any state missing some valid Markov transitions. Those transitions either point back to the original Markov automaton, for Markov transitions that do not start any no-good, or point to the states of the first layer of the trie, for Markov transitions that start a new no-good. Since we kept the transitions to the non-accepting states that complete a no-good, we know we are not introducing any no-good. Now, we can finally remove those non-accepting states (line 33). It is important to observe that all the steps we described maintain the invariant that for each state of the automaton being built, the label of all its incoming transitions is always the same. This property will be exploited for sampling the sequences recognised by the automaton.

The algorithm adds exactly once each transition of the resulting automaton. Therefore, it runs in time linear in the number T of transitions of the final automaton. Let $N = L \cdot |\mathcal{N}|$, where L is the size of a no-good, be the size of the input $\mathcal{N}$. When constructing the trie, it creates exactly N transitions. During the next phase, the added transitions are exactly those added by the Aho and Corasick algorithm. Their number is linearly bounded by N, a (non-trivial) result from Aho and Corasick [1]. Finally, the number of transitions added to each state during the completion phase is bounded by $|\Sigma|$, which is a constant independent of $\mathcal{N}$.

Note that the general idea of this algorithm is similar to the algorithm by [14], which computes shortest paths with forbidden paths. However, they operate in a very different context, and are only interested in the shortest paths of a graph, whereas we are interested in all the sequences accepted by the original Markov automaton.

Algorithm 1: Computing the no-good automaton

Data: $\mathcal{N}$ the set of forbidden no-goods
$\mathcal{M} \leftarrow \langle Q, \Sigma, \delta, q_0, F \rangle$: a Markov automaton
Result: Any word containing at least one no-good is not recognised by $\mathcal{M}$

```
    // Remove transitions that start a no-good
 1  forall the a₁ ... aₗ ∈ N do
 2  │   q ← Q(a₁)
 3  │   Clear outgoing transitions of q
 4  └   w(q) ← (a₁)

    // Compute the trie of no-goods
 5  Q_trie ← ∅
 6  forall the a₁ ... aₗ ∈ N do
 7  │   q ← q₀
 8  │   i ← 1
 9  │   while δ(q, aᵢ) exists do
10  │   │   q ← δ(q, aᵢ)
11  │   └   i ← i + 1
12  │   for aⱼ, i ≤ j ≤ L do
13  │   │   q' ← NewState(Q_trie)
14  │   │   F ← F ∪ {q'}
15  │   │   δ(q, aⱼ) ← q'
16  │   │   w(q') ← (a₁, ..., aⱼ)
17  │   │   a(q') ← {aⱼ}
18  │   └   q ← q'
19  └   F ← F \ {q}

    // Compute cross prefix transitions
20  forall the q ∈ Q_trie do
21  └   S(q) ← {q' ∈ Q_trie | w(q) is a strict suffix of w(q')}

22  forall the q ∈ Q_trie in order of decreasing |w(q)| do
23  │   forall the a ∈ Σ such that δ(q, a) exists do
24  │   │   forall the q' ∈ S(q) do
25  │   │   │   if δ(q', a) is undefined then
26  │   │   │   └   δ(q', a) ← δ(q, a) // transition is Markovian
27  │   │   │
    │   │   └
    │   └
    └

    // Markovian completion
28  forall the ∀q ∈ Q_trie do
29  │   {a₁} ← a(q)
30  │   forall the a₂ ∈ Σ such that a₁a₂ ∈ C do
31  │   │   if δ(q, a₂) is undefined then
32  │   │   └   δ(q, a₂) ← Q(a₂)
    │   └
    └
33  Q ← Q ∪ (Q_trie ∩ F)
```

4 Sampling Sequences with a Maximum Order Guarantee

We now describe how to sample Markov sequences containing no forbidden no-good, with its correct probability, using the no-good automaton computed in the previous section. As a first approximation, one can use the no-good automaton to generate new sequences, since all walks in this automaton produce valid sequences. The transition probabilities of the original Markov model are used for walking in this automaton, by choosing the successor of a state q as follows: with $a(q)$ the label of all transitions pointing to q, choose a transition labelled with b with probability $P(b|a(q))$. We show in the next section that this produces a crude approximation of the probability of sampled sequences. However, if we impose a given sequence length, we can use *belief propagation* and do perfect sampling, by exploiting a special case where belief propagation is, by principle, exact and polynomial.

Belief propagation [11] is an algorithm, or family of algorithms, for performing statistical inference on graphical models, i.e. graph-based representations of distributions over multiple variables. Specifically, we consider graphical models called *factor graphs*. A factor graph is a bipartite undirected graph, representing the factorisation of a probability function, where nodes represent either variables or factors, and edges connect factors to the variables to which that factor applies. The sum–product algorithm allows us to perform statistical inference, such as marginalising or sampling the probability function.

4.1 Background on Belief Propagation

Suppose we have n random variables $X_1, \ldots, X_n$, and a real-valued function $g(X_1, \ldots, X_n)$ that can be expressed as the product of m factors:

$$g(X_1, \ldots, X_n) = \prod_{j=1}^{m} f_j(S_j),$$

where the factor f_j is a function holding only on a subset $S_j \subseteq \{X_1, \ldots, X_n\}$ of the variables. The corresponding factor graph is a bipartite graph $G = (X, F, E)$, where $X = \{X_1, \ldots, X_n\}$, $F = \{f_1, \ldots, f_m\}$, and an edge (X_i, f_j) is in E iff $X_i \in S_j$.

Example 1 Consider a function holding on three variables X_1, X_2, X_3, defined as the product of four factors:

$$g(X_1, X_2, X_3) = f_1(X_1, X_2) \cdot f_2(X_2, X_3) \cdot f_3(X_1, X_3) \cdot f_4(X_3)$$

Figure 4 shows the corresponding factor graph.

In general, functions that display a tree-structured factor graph have interesting properties, since several statistical properties can be computed in an exact

Fig. 4 The factor graph for
the function
$g(X_1, X_2, X_3) =$
$f_1(X_1, X_2) \cdot f_2(X_2, X_3) \cdot$
$f_3(X_1, X_3) \cdot f_4(X_3)$

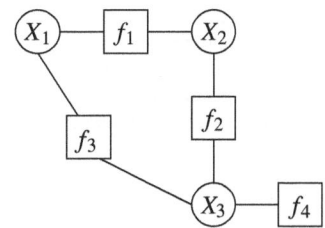

and yet tractable way, by exploiting the independence between subsets of variables. To illustrate this on our previous example, suppose factor f_3 is suppressed, i.e. $g(X_1, X_2, X_3) = f_1(X_1, X_2) \cdot f_2(X_2, X_3) \cdot f_4(X_3)$. The resulting factor graph thus becomes a tree. Suppose we want to compute the normalisation constant Z such that $P(X_1, X_2, X_3) = \frac{1}{Z} \cdot g(X_1, X_2, X_3)$ is a probability function. In general, $Z = \sum_{X_1} \sum_{X_2} \sum_{X_3} g(X_1, X_2, X_3)$, implying a summation over the whole cartesian product of the domains of the variables of g, which is exponential in the number of variables. However, by exploiting the tree structure of the factor graph, we can first compute $\sum_{X_3} f_2(X_2, X_3) \cdot f_4(X_3)$, which results in a function $f_{X_2}(X_2)$ on X_2 only, and then we can compute the sum $\sum_{X_1, X_2} f_1(X_1, X_2) \cdot f_{X_2}(X_2)$. In other words, we only perform local summations at a time, on the cartesian product of the variables of a given factor, and never on the whole set of variables.

The sum–product algorithm, first invented in 1982 by Pearl [10], is an algorithm that computes the marginal function of a particular node, i.e. the function $g(X_i) = \sum_{X_1,...,X_{i-1},X_{i+1},...,X_n} g(X_1, \ldots, X_n)$ for some X_i, using the same strategy. Likewise, if the factor graph is a tree, it gives the exact result in tractable time. This algorithm is often referred to as message passing. In turn, this can be used to sample valuations of $X_1, \ldots, X_n$ with their right probability. One needs to draw values for all X_i, from X_1 to X_n, choosing each time a value for X_i according to its marginal probability $g(X_i)$.

4.2 A Factor Graph Model of Max Order Sequences

We now apply those techniques to our application, and describe message passing, by instantiating this procedure to the specific problem of sampling Markov sequences containing no forbidden no-good of size L. The problem of generating a Markov sequence of n variables $X_1, \ldots, X_n$ containing no forbidden no-good, is the problem of sampling the function g_{maxo} defined as:

$$g_{maxo}(X_1, \ldots, X_n) = \begin{cases} 0 & \text{if } X_1, \ldots, X_n \text{ contains a no-good,} \\ \frac{1}{Z} \cdot P(X_1) \cdots P(X_n | X_{n-1}) & \text{otherwise} \end{cases}$$

Fig. 5 The basic factor graph for max order sequences, with a Markov order $k = 1$, a maximum order $L = 3$ and a length of $n = 5$

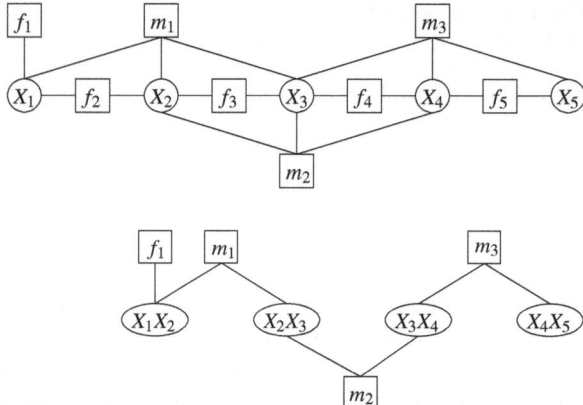

Fig. 6 The—inefficient—acyclic reformulation of the factor graph of Fig. 5

This function g_{maxo} can be represented in a straightforward way with a factor graph, such as the one shown on Fig. 5, for the case where $k = 1$, $L = 3$ and $n = 5$. We have $f_1(X_1) = \frac{1}{Z} \cdot P(X_1)$, $f_i(X_{i-1}, X_i) = P(X_i|X_{i-1})$, and $m_i(X_i, \ldots, X_{i+L-1}) = 0$, if $X_i, \ldots, X_{i+L-1}$ is a no-good of size L, and $m_i(X_i, \ldots, X_{i+L-1}) = 1$ otherwise.

As it stands, this factor graph is not acyclic. In order to render it acyclic, one can apply a naive, "brute force", reformulation: we merge every subsequence of $L - 1$ consecutive variables into one single merged variable. As a result, we have only binary factors between every two consecutive merged variables, as shown on Fig. 6. When those variables are assigned two overlapping $(L - 1)$-grams that combine into an L-gram, and that this L-gram is not a no-good, the binary factor gives the Markov probability of the L-gram; otherwise it evaluates to 0. However, this would not be tractable in practice, since the alphabets of the merged variables contain all combinations of $L - 1$ elements of the original alphabet, and this is exponential in the max order L.

We propose another reformulation, which exploits the no-good automaton. We define a function $g(X_1, \ldots, X_n)$, where the domain of each variable X_i is the set of states of the no-good automaton, and not the states of the original Markov model. Recall that the states of the original Markov chain correspond to the set Σ of labels of the no-good automaton. This function g is composed of simple binary factors, and its factor graph decomposition is shown on Fig. 7.

Fig. 7 The factor graph for max order using the no-good automaton

In order to define the factors involved in this factor graph, we exploit the fact that, for each state q of the automaton, the label of all incoming transitions is the same, denoted $a(q)$. As we established earlier, this is satisfied both for the Markov automaton and for the no-good automaton produced by our algorithm. The same binary factor is applied along the sequence, i.e. $\forall i \leq n$, $f_i = f$, with f defined as follows:

$$f(q, q') = \begin{cases} P(a(q')|a(q)), & \text{if } q' \in \delta^+(q) \text{ and } q \neq q_0, \\ 0 & \text{otherwise} \end{cases}$$

This binary factor imposes that sequences are formed by walking the automaton, with a probability given by the Markov probability. The unary factors impose that sequences are formed by traversing the automaton from the initial state to an accepting state, and can additionally impose a bias on the element at a given position of a solution. Suppose $P_i(X_i)$ is a probability distribution on Σ, biasing the probability of the Markov state occurring at position i, such that the probability of a sequence $X_1, \ldots, X_n$ is the biased Markov probability $P(X_2|X_1).\ldots. P(X_n|X_{n-1}).P_1(X_1).\ldots.P_n(X_n)$. Typically, $P_1(X_1)$ imposes the prior probability on X_1. Hard unary constraints, where some values are forbidden, and the remaining ones have a uniform probability, can also be imposed under this formalism. The unary factors of our model are thus defined as follows:

$$m_i(q) = \begin{cases} 0, & \text{if } q = q_0, \\ P_i(a(q)), & \text{otherwise} \end{cases}$$

To this general definition, we need to add special cases for $i = 1$ and $i = n$. For $i = 1$ we introduce to the equation the additional case $m_1(q) = 0$ if $q \notin \delta^+(q_0)$, stating that we can only start a sequence from a successor of the initial state of the automaton. For $i = n$, we introduce the case $m_n(q) = 0$ if $q \notin F$, stating that we can only end a sequence on an accepting state of the automaton.

It is easy to verity that $g(X_1, \ldots, X_n) = g_{maxo}(a(X_1), \ldots, a(X_n))$, if we do not take into account the unary factors m_i, $i > 1$. Additionally, since each sequence of Markov states correspond to a unique sequence of states in the automaton from the initial state to an accepting state, each sequence of Markov states corresponds to a unique sequence of states $X_1, \ldots, X_n$ with a non-zero g probability. As a result, sampling g_{maxo} is equivalent to sampling g. Note that if we incorporate the additional unary factors m_i, $i > 1$ into the definition of g_{maxo} too, the exact same reasoning holds.

4.3 Sampling Max Order Sequences

Algorithm 2 is a description of the sum–product algorithm applied specifically to our factor graph, and specialised for sampling sequences. The sum–product algorithm is used to compute marginalisations, which in turn is used to sample solutions with their correct probability, in the following way. Let $g_{X_1}(X_1)$ be the marginalisation of g on X_1, equal to $\sum_{X_i, i>1} g(X_1, \ldots, X_n)$. Intuitively, for a given value q_1 of X_1, $g_{X_1}(q_1)$ gives us the probability that $X_1 = q_1$ in a sequence of values of $X_1, \ldots, X_n$, drawn randomly according to this sequence's probability. Suppose we have drawn $X_1 = q_1$ with this probability, $g(q_1, X_2, \ldots, X_n)$ is now a function of only $X_2, \ldots, X_n$. If we marginalise this function on X_2, we get a function $g_{X_2}(X_2) = \sum_{X_i, i>2} g(q_1, X_2, \ldots, X_n)$, giving the probability $g_{X_2}(q_2)$ that $X_2 = q_2$ in a sequence of values of $X_1, \ldots, X_n$ starting with $X_1 = q_1$, drawn randomly according to its g probability. After repeating this process until all X_i have been assigned, we end up with a sequence of values $q_1, \ldots, q_n$, drawn with the probability that $X_1 = q_1$ and $X_2 = q_2$ and ... and $X_n = q_n$, which is exactly the probability of the valuation $q_1, \ldots, q_n$ in the distribution g.

On the specific case where the factor graph is a tree, as in the model of Fig. 7, the sum–product algorithm performs in two phases only. A first phase from the leaf to the root (for any arbitrarily chosen root) of the tree, and a second from the root to the leaf. In our case, it is convenient to traverse the graph from X_n down to X_1 (backward phase), and from X_1 back to X_n (forward phase).

During the *backward phase*, at the ith iteration, the algorithm considers only the part of the factor graph from X_i to X_n, i.e. the function $g_{i \leftarrow}(X_i, \ldots, X_n) = m_i(X_i).f_i(X_i, X_{i+1}).\cdots.f_{n-1}(X_{n-1}, X_n).m_n(X_n)$. It computes a *message* on X_i, which is simply a probability distribution $g_{i \leftarrow}(X_i)$ on X_i, equal to the marginalisation on X_i of $g_{i \leftarrow}(X_i, \ldots, X_n)$. At the end of the backward phase, the message on X_1 is the normalisation of the full function g on X_1, i.e. $g_{1 \leftarrow}(X_1) = g_{X_1}(X_1)$.

At the beginning of the *forward phase*, the algorithm draws a value for X_1 according to $g_{X_1}(X_1)$. Then, at the ith iteration of the forward phase, the algorithm has already instantiated $X_1, \ldots, X_{i-1}$ to $q_1, \ldots, q_{i-1}$. It considers the part of the factor graph from X_1 to X_i, i.e. the function $g_{i \rightarrow}(X_1, \ldots, X_i) = m_1(X_1).f_1(X_1, X_2).\cdots.f_{i-1}(X_{i-1}, X_i)$. The message $g_{i \rightarrow}(X_i)$ on X_i is essentially the marginalisation of $g_{i \rightarrow}(q_1, \ldots, q_{i-1}, X_i)$. Since the product $g_{i \rightarrow}(X_1, \ldots, X_i).g_{i \leftarrow}(X_i, \ldots, X_n)$ is equal to $g(X_1, \ldots, X_n)$, for a value q_i of X_i, the product $g_{i \rightarrow}(q_i).g_{i \leftarrow}(q_i)$ is precisely the probability $g_{X_i}(q_i)$, i.e. the probability that $X_i = q_i$ in a sequence on $X_1, \ldots, X_n$ starting with $q_1, \ldots, q_{i-1}$, drawn randomly with probability given by g. Therefore, the algorithm draws q_i with probability $g_{i \rightarrow}(q_i).g_{i \leftarrow}(q_i)$, and continues up to iteration n, when all variables are instantiated. Note that when sampling several sequences, the backward phase needs to be performed only once, and the forward phase will sample a new sequence every time with its correct probability.

5   Evaluation

Following our opening example, we applied our algorithm on the "Eugene Onegin" corpus (2160 sentences, 6919 unique words, 32,719 words in total). On this corpus, depending on the Markov order (ranging from 1 to 3) and the maximum order

---

Algorithm 2: Sum–product algorithm for max order sampling

Data: Function $g(X_1, \ldots, X_n)$ and its factor graph
Result: A sequence $q_1, \ldots, q_n$, with probability $g(q_1, \ldots, q_n)$

```
// Backward phase
```
$\mathbf{g}_{n\leftarrow}(X_n) \leftarrow m_n(X_n)$
for $i \leftarrow n - 1$ *to 1* **do**
    foreach $q \in Q$ **do**
        $\mathbf{g}_{i\leftarrow}(q) \leftarrow \sum_{q' \in Q} m_i(q).f_i(q, q').\mathbf{g}_{i+1\leftarrow}(q')$
    Renormalise $\mathbf{g}_{i\leftarrow}$

```
// Forward phase
```
$\mathbf{q_1} \leftarrow draw(\mathbf{g}_{1\leftarrow}(X_1))$
for $i \leftarrow 2$ *to n* **do**
    foreach $q \in Q$ **do**
        $\mathbf{g}_{i\rightarrow}(q) \leftarrow f_{i-1}(\mathbf{q_{i-1}}, q)$
    Renormalise $\mathbf{g}_{i\rightarrow}$
    $\mathbf{q_i} \leftarrow draw(\mathbf{g}_{i\rightarrow}(X_i).\mathbf{g}_{i\leftarrow}(X_i))$
return $(\mathbf{q_1}, \ldots, \mathbf{q_n})$

---

parameters (ranging from 3 to 20), which both affect the number and the size of no-goods, computing the no-good automaton takes consistently around 200 ms and never more than 300 ms.

5.1 Solution Loss

An interesting question is how likely a stochastic method finds sequences satisfying the maximum order property. In order to assess this, we try to estimate the probability that a Markov sequence contain no no-good. We perform this estimation by making simple solution counting: we count the number of sequences of a given length with a non-zero probability, with and without a max order constraint, and consider the ratio of the two values. Under the assumption that valid max order sequences follow a distribution similar to that of all Markov sequences, this ratio should be a good estimation of the probability of satisfying the imposed max order. We first count the total number S of Markovian sequences of length $n = 20$, with a Markov order 3, based on the Eugene Onegin corpus. This is obtained by raising to the power of 20 the transition matrix where non-zero entries are replaced with 1. We compare this to the number S_L of Markovian sequences with a maximum order of L, with L ranging from 5 (the minimum non-trivially infeasible value), to 20 (the maximum value for which max order is not trivially satisfied). This is obtained by raising to the power of 20 the adjacency matrix of the no-good automaton (i.e. the matrix where 1-entries correspond to pairs of linked states). We call *solution loss* the ratio $1 - (S_L/S)$: the closer it is to 1, the more Markovian sequences are "ruled out" because they do not satisfy the maximum order property for L. We show the results on Fig. 8. Naturally,

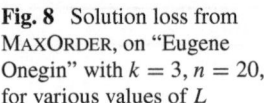

Fig. 8 Solution loss from
MAXORDER, on "Eugene
Onegin" with $k = 3$, $n = 20$,
for various values of L

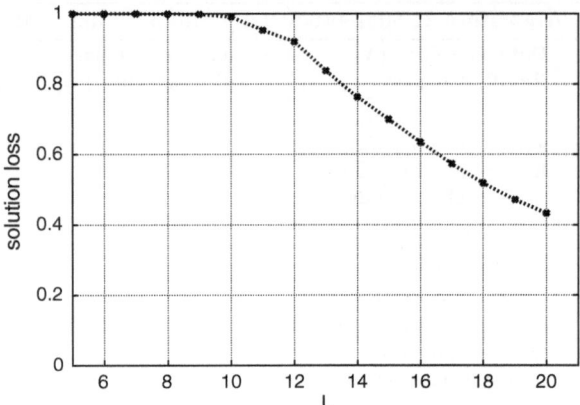

the constraint is tighter for low values of L. For $L = 5$ for example, there is no
solution, leading to a solution loss of 1. For bigger values, the solution loss is still
close to 1, with less that 1 % solutions left.

Note that a formal combinatorics approach for characterising this ratio has been
given by Guibas and Odlyzko [6], where they define a generating function for the
number of sequences of a given length not containing any forbidden pattern from a
list of forbidden patterns. This generating function has a closed form, which can be
used, in principle, to estimate the number of such sequences using partial fraction
decompositions. However, it is much more practical, and in principle equivalent, to
estimate this number based on the no-good automaton, as we did here.

5.2 Sampling

We compare the probability of generating a sequence by the two sampling methods
mentioned in Sect. 4: a random walk in the no-good automaton, and our fixed-length
belief propagation model. The purpose of this experiment is twofold: show that a
random walk in the automaton does not sample sequences correctly, and confirm
empirically that our belief propagation-based model is indeed correct.

We applied each method to generate sequences of length 8, of Markov order 1 and
with an imposed max order of 4. For the random walk method, we imposed the length
simply by rejecting shorter sequences. In total, we sampled over 20 million sequences.
Of those, 5 million were unique sequences. Concerning running times, the baseline
random walk-based procedure generated an average of 5500 sequences per second
(counting only non-rejected sequences), while the exact belief propagation-based
method generated an average of 3500 sequences per second. To measure empirical

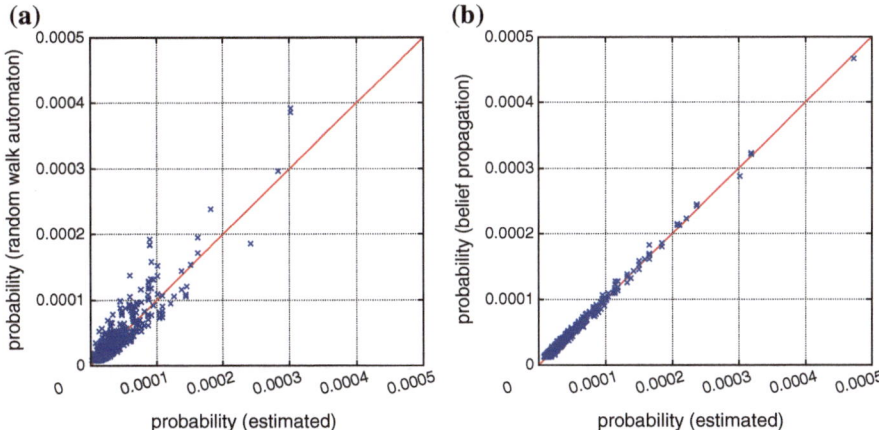

Fig. 9 Sampling with random walk in the automaton compared to belief propagation. A point corresponds to a sequence, having a certain probability (*x*-axis), sampled with a certain frequency (*y*-axis). **a** Random walk in the automaton. **b** Belief propagation

probabilities more accurately, we filtered only those that were generated over 50 times, of which there were about 47,000 with random walk, about 35,000 with belief propagation (because of the slight time penalty for belief propagation). We used a simple method to estimate the correct probability of a max order sequence, by computing Z, equal to the sum of the probability of all unique sequences found by either method, and use $1/Z$ as the normalising constant, i.e. the probability of a max order sequence is equal to its Markov probability divided by Z. With the high number of sequences that we generated, this gave a reasonably accurate estimation.

We plot our results on Fig. 9. Each point on either graph corresponds to a sequence. Its value on the *x*-axis is its actual probability, estimated as described previously, while the values on the *y*-axis is the empirical probability, i.e. the frequency at which the specific sequence has been sampled compared to the total number of sequences. Figure 9a shows that the baseline sampling approach performs rather poorly: many sequences, even of similar probability, are over or under-represented. On the other hand, Fig. 9b provides a striking empirical confirmation of the correctness of the belief propagation model.

5.3 Example

We conclude this section by showing an example of a text generated with this method, with a Markov order 2 and a maximum order 6:

Look to the circle of our first ages from thirty down to the end . He's moved . For cousins from afar darlings then we'll throw at him . Never . She was still helping the poor butterfly. Happy is he apparelled . Is this the man of honour and the marriage-bed, in all the play of hope ? He failed to understand and took deep in gloom and mist . I beseech, and take a swill . He arrives, the girl's attentive eyes are dreaming . But to the bereaved, as if beneath her pillow, his father died . From her husband's or the unaffected thoughts of all that is the advent of the hall .

By construction, chunk sizes are bounded by 6. For information, about 49 % of the chunks of the sequence were of size 5, 32 % of size 4, and 19 % of size 3. The max order guarantee implies that no copy of size 6 or more is made from the corpus.

6 Conclusion

We have introduced the problem of generating Markov sequences satisfying a maximum order, or more generally, containing no forbidden sequence (no-good), an important issue with Markov models that has not, to our knowledge, been addressed previously. In the first step, we formulate the problem in the framework of automata theory, and exhibit an automaton that solves the algebraic problem of generating Markov sequences with no no-goods. In the second step, we extend the automaton and associate it with a linear factor graph to achieve perfect sampling of these sequences, thereby closing the issue. The set of no-goods can be arbitrary (finite) so as to encode any unwanted set of subsequences (not only coming from the corpus). Interestingly, the inverse problem (*plagiaristic generation*) consisting in guaranteeing that at least one no-good is present in each generated sequence is not of the same nature. This is work in progress.

References

1. Aho, A.V., Corasick, M.J.: Efficient string matching: an aid to bibliographic search. Commun. ACM **18**(6), 333–340 (1975)
2. Begleiter, R., El-Yaniv, R., Yona, G.: On prediction using variable order markov models. J. Artif. Intell. Res. (JAIR) **22**, 385–421 (2004)

3. Beldiceanu, N., Carlsson, M., Petit, T.: Deriving filtering algorithms from constraint checkers. In: [15], pp. 107–122
4. Brooks, F.P., Hopkins, A., Neumann, P.G., Wright, W.: An experiment in musical composition. IRE Trans. Electron. Comput. **6**(3), 175–182 (1957)
5. Conklin, D., Weisser, S.: Antipattern discovery in ethiopian bagana songs. In: Dzeroski, S., Panov, P., Kocev, D., Todorovski, L. (eds.) Discovery Science - 17th International Conference, DS 2014, Bled, Slovenia, 8–10 October 2014. Proceedings. Lecture Notes in Computer Science, pp. 62–72. Springer (2014)
6. Guibas, L.J., Odlyzko, A.M.: String overlaps, pattern matching, and nontransitive games. J. Comb. Theory, Ser. A **30**(2), 183–208 (1981). doi:10.1016/0097-3165(81)90005-4
7. Karakostas, G., Lipton, R.J., Viglas, A.: On the complexity of intersecting finite state automata. IEEE Conference on Computational Complexity, pp. 229–234. IEEE Computer Society (2000)
8. Karakostas, G., Lipton, R.J., Viglas, A.: On the complexity of intersecting finite state automata and NL versus NP. Theor. Comput. Sci. **302**(1–3), 257–274 (2003)
9. Papadopoulos, A., Roy, P., Pachet, F.: Avoiding Plagiarism in Markov Sequence Generation. In: Brodley, C.E., Stone, P. (eds.) AAAI. AAAI Press, Menlo Park (2014)
10. Pearl, J.: Reverend bayes on inference engines: a distributed hierarchical approach. In: Waltz, D.L. (ed.) Proceedings of the National Conference on Artificial Intelligence. Pittsburgh, PA, 18–20 August 1982, pp. 133–136. AAAI Press (1982)
11. Pearl, J.: Probabilistic reasoning in intelligent systems - networks of plausible inference. Morgan Kaufmann series in representation and reasoning. Morgan Kaufmann (1989)
12. Pesant, G.: A regular language membership constraint for finite sequences of variables. In: [15], pp. 482–495
13. Pinkerton, R.C.: Information theory and melody. Scientific American (1956)
14. Villeneuve, D., Desaulniers, G.: The shortest path problem with forbidden paths. Eur. J. Oper. Res. **165**(1), 97–107 (2005)
15. Wallace, M. (ed.): Principles and Practice of Constraint Programming - CP 2004. 10th International Conference, CP 2004, Toronto, Canada, 27 September–1 October 2004, Proceedings. Lecture Notes in Computer Science. Springer (2004)

Integrating Purpose and Revision into a Computational Model of Literary Generation

Pablo Gervás and Carlos León

Abstract Over the past few years, advances in the area of computational creativity have explored the combination of generative processes and evaluation models to obtain artifacts that are both original and valuable. This has taken place in fields as different as music, graphical art, or mathematics. Many of these efforts rely on identifying a sweet spot where an existing technology provides a constructive process that resembles closely some part of a creative process of humans. However, these approaches very rarely model the iterative nature of creative process as observed in humans, where a creator sets out with a purpose in mind, and creates drafts and revises them successively until the purpose is met. While it is a fact that in truly creative processes the purpose may also evolve during revision, the lack of direction has always been a damaging criticism to early approaches to computational creativity. In this chapter we will address a set of examples of approaches to the computational generation of literary texts based on particular techniques, propose a computational model that captures the purpose-driven revision of generated artifacts, and review a number of ongoing research efforts that implement parts of this model.

1 Introduction

The goal of modelling the ability of humans to produce literary artifacts has been a long-standing dream of artificial intelligence. Poetry generation and story telling were among the earliest tentative applications of computers to modelling human behaviour.

P. Gervás (✉)
Instituto de Tecnología del Conocimiento, Universidad Complutense de Madrid,
Madrid, Spain
e-mail: pgervas@ucm.es

C. León
Departamento de Ingeniería del Software e Inteligencia Artificial,
Universidad Complutense de Madrid, Madrid, Spain
e-mail: cleon@ucm.es

© Springer International Publishing Switzerland 2016
M. Degli Esposti et al. (eds.), *Creativity and Universality in Language*,
Lecture Notes in Morphogenesis, DOI 10.1007/978-3-319-24403-7_7

This interest by pioneer researchers in computing and AI in the generation has been described for poetry [2] and for storytelling [21]. These efforts were usually based on the exploration of elementary processes or techniques that had become available, and how they might be applied to the tasks under consideration. As such, they usually focused on taking advantage of how specific aspects of the task in question happened to be captured by the technique being used, rather than trying to establish how the underlying complexity of the task itself should be modelled for an optimal solution to the problem. Many of these efforts rely on identifying a sweet spot, where an existing technology provides a constructive process that resembles closely some part of a creative process of humans. However, these approaches very rarely model the iterative nature of creative process as observed in humans, where a creator sets out with a purpose in mind, and creates drafts and revises them successively until the purpose is met. While it is a fact that in truly creative processes the purpose may also evolve during revision, the lack of direction has always been a damaging criticism to early approaches to computational creativity.

The consideration of purpose in approaches to computational creativity has never gone far beyond a generic goal for the artifacts under construction to be useful or to fit into an overall genre. Such generic criteria have been captured formally by Ritchie [47] in his ground-breaking paper on the evaluation of computational creative systems. Part of the difficulty in addressing purpose lies in the fact that the conceptual spaces to be searched are very large and the density of acceptable solutions to be found in them is very low even when a simple specification of the desired form is used to drive the process. To impose additional constraints on the output related to purpose beyond aesthetics runs the risk of ruling out the most satisfying solutions from the point of view of form. Yet this optimality of both form and purpose is a fundamental ingredient in the perception of creativity, and as such, every effort should be made to include it in computational models of the corresponding tasks.

To this end, it may be important to start from a theoretical model of the process that takes into consideration the following points:

- the constructive process must start from a specification of the desired output
- the process should take such specification into account to drive the construction, but may also need to modify this specification so that it leads to more successful areas of the conceptual space being explored
- the initial specification may be revised if this results in outputs that exceed the expectations of the original purpose

In this chapter, we will address a set of examples of approaches to the computational generation of literary texts based on particular techniques, and describe a computational model of the creative process for literary texts—the ICTIVS model [28]—that captures the purpose-driven revision of generated artifacts in a way that satisfies these restrictions. To illustrate the model, we then review a number of ongoing research efforts that implement parts of this model. Finally, we close the chapter with some conclusions arising from the material presented.

2 Computational and Cognitive Models of the Writing Task

This section reviews a number of models of the writing task, both from a computational point of view—efforts in automated poetry generation and automated generation of narrative—and from a cognitive point of view—attempts to understand the processes that humans apply to the writing task.

2.1 Poetry Generation

By its nature, the task of generating a poem, when addressed by either a computer or a human, has to satisfy constraints at two very different levels. One level concerns the sequence in which the words appear in the poem. For a draft to be acceptable there has to be some way in which the words in it appear to link to one another, to make sense as a linguistic message. This constraint is applicable to the whole poem but essentially it operates at a local level, based on how each word can be seen to follow on from the previous one. A different level concerns certain macro-structural features that may be desirable in a poem, such as being distributed over a number of lines of specific lengths in terms of syllables, or having rhyming words occur at the end of particular lines. This corresponds to the poem satisfying some form of poetic stanza.

Computer generation of poetry has traditionally addressed these constraints in terms of two different strategies: one is to reuse large fragments of text already formatted into poem-like structures of lines, and the other is to generate a stream of text by some procedure that ensures word-to-word continuity and then establish a distribution of the resulting text into lines by some additional procedure.

The reuse of text fragments already distributed into poetic lines was pioneered by [41, 45] and it has more recently been used by [9, 10, 29, 48, 52, 54, 56]. In all these cases, either lines or larger poem fragments from existing poems are subjected to modifications—usually replacement of some of the words with new ones—to produce new poems. In a refinement on this method, the selected fragment is stripped down to a skeleton consisting only of the POS tags of each line, and words corresponding to the desired content are used to fill this skeleton in. This procedure is followed in [1, 20, 53].

Alternative procedures rely on building a stream of text from scratch, and resort to various techniques to ensure the continuity of the textual sequence. One early approach was to rely on linguistic grammars to drive the construction. This was the approach followed in [38, 39], where TAG grammars were employed. A more popular alternative is the use of n-grams to model the probability of certain words following on from others. This corresponds to reusing fragments of the corpus of size n, and combining them into larger fragments based on the probability of the resulting sequence. This is the main approach for ensuring text coherence used in

[6, 14, 23, 25]. All these different computer poets rely on various additional methods for establishing constraints on the resulting poem drafts.

To ensure that resulting poems satisfy constraints on poem structure in terms of lines, systems that build a stream of text from scratch rely on either building each line separately [14] or applying a separate procedure for distributing the resulting text into poetic lines [23, 25].

2.2 Narrative Generation

In the case of narrative generation, existing efforts have focused on two different aspects of the task. Earlier efforts focused on the task of coming up with the underlying argument of the story, with little concern over its final form as text. More recent efforts have started to focus on the way in which a given story can be told. This distinction matches closely that established by narratologists between story—what is told, also known as *fabula*—and discourse—the way in which it is told, also knows as *sjuzet*. We review these two sets of efforts separately.

2.2.1 Building Arguments for Stories Automatically

The automated construction of story arguments has drawn from two different sources: existing AI techniques, and existing studies of narrative. The most popular AI technique employed for storytelling has been planning. Efforts along these lines rely on interpreting a story as a plan, and using AI planning techniques to generate a path—a set of chained actions—that lead from an initial situation to a desired goal. Approaches to storytelling based on narratology have generally favoured the work of structuralists, who analysed narrative in terms of its structure. Based on these analyses of the structure of narrative, solutions have been put forward to generate artifacts with corresponding structures. Both approaches are reviewed briefly below.

It is clear that planning has been central to efforts of modelling storytelling for a long time. Most of the existing storytelling systems [21] feature a planner of some kind, whether as a main module or as an auxiliary one. The set of events to be included in a narrative are generated as the solution to one or several planning problems. This ensures that all the events in the resulting narrative are, by construction, linked by causal chains. The planner in TALESPIN [40] addressed the issues of decomposition of a goal into subgoals and of explicitly representing character goals. Each of these issues has later been the focus of specific research efforts on planning for narrative. The IPOCL planner used by Fabulist [49] combines causal reasoning with a simulated intention recognition process to drive plan creation. This allows the construction of plans where character actions seem to be motivated by agents' internal beliefs and desires, which leads to a positive impression of character believability. The rules employed by Novel Writer [33] operated basically as patterns for fragments of story plots which were instantiated by the program and used to construct an overall plot.

The plot fragments used by AUTHOR [13] follow very similar lines (though they extend Novel Writer's rules by including explicit author goals that can lead to recursive creation of a complex structure of the plot in terms of how the author goals are resolved).

MINSTREL [55] combined both the author goals and the character goals by introducing two different modules: the planner (for author goals) and the problem solver (for character goals).

Of the many theories of narrative developed in the Humanities, only a few have bridged the gap to become tools in the hands of AI researchers. Of these, Propp's Morphology of the Folktale [44], is the most extended one, having been applied in several AI systems for story generation. The two corner stones of Propp's analysis of Russian folk tales are a set of roles for characters in the narrative (which he refers to as *dramatis personae*), and a set of character functions, understood as acts of the character, defined from the point of view of its significance for the course of the action. These have been used in several systems [15, 17, 18, 31, 55, 58], represented in slightly different ways. Of all these, the most explicit conceptual representation of Propp's set of narratological categories is the description logic formulation developed by Peinado [43], where these concepts are defined as an OWL ontology. Propp's manner of abstracting narrative structure from a set of stories is far from being the only possible one. At a different level of detail, another favourite is the three-act restorative structure. This model derived from Joseph Campbell's analysis of the structure of myths [8], which is a dominant formula for structuring narrative in commercial cinema [57]. Another source that is also being considered in AI is the work of Chatman [11]. This model constitutes a step up from the models of Propp or Campbell in the sense it considers a wider range of media, from literature to film. Chatman studies the distinction between *story* and *discourse*, and he proposes ways of decomposing each of these domains into elementary units.

2.2.2 Deciding Automatically How to Tell a Story

The task of composing a narrative based on a given set of events that have taken place has received little attention in terms of conceptual modelling. Efforts have been made to capture the structure of narratives as a finished product (by the narratology research community [19]), to come up with a set of cognitive processes implied in the tasks of writing in general or of understanding narrative in particular (by the cognitive science community [16, 51]), to build models of how discourse may be structured for a given plot (by the artificial intelligence community [7, 32]) and to construct functional architectures for generating text from conceptual data (by the natural language generation community [3, 4, 34]). The task of putting together a narrative that conveys events that have already happened is related to all these aspects. It is also the kind of basic storytelling that people carry out in their everyday lives to communicate with one another, to convince, to inform, to remember the past, to interpret the present and to plan for the future. Efforts in AI on structuring appropriate discourse for conveying a given plot include work on use of flashback and foreshadowing to produce surprise [7] and automatic generation of camera placements over time to define a visual discourse that best fits the plot to be rendered [32]. Both of these efforts rely on a planning-based approach to narrative, with plots represented as plans. A number of related efforts exist to automatically derive narratives from sport games [3, 4, 34]. These efforts operate on input data in the form of statistics on a given game, and produce texts in the manner of newspaper articles covering similar games. The task of narrative composition has also been explicitly addressed by Hassan et al. [30], and Gervás [22, 24]. Hassan et al. addressed the task of generating stories from the logs of a social simulation system. Gervás focused on the task of generating a natural language rendering of a story extracted from a chess game for a given set of pieces.

2.3   Cognitive Accounts of Writing

The human ability to compose understandable and successful text—both in terms of communicative purpose and aesthetics—has been the subject of many studies. Two of them are particularly relevant to the points being made in this chapter.

Flower and Hayes [16] define a cognitive model of writing in terms of three basic process: planning, translating these ideas into text and reviewing the result with a view to improving it. These three processes are said to operate interactively, guided by a monitor that activates one or the other as needed. The planning process involves generating ideas, but also setting goals that can later be taken into account by all the other processes. The translating process involves putting ideas into words, and implies dealing with the restrictions and resources presented by the language to be employed. The reviewing process involves evaluating the text produced so far and revising it in accordance to the result of the evaluation. An important feature to be considered is that the complete model is framed by what Flower and Hayes consider "the rhetorical problem", constituted by the rhetorical situation, the audience and the writer's goals.

Sharples [50] presents a description of writing understood as a problem-solving process where the writer alternates between the simple task of exploring the conceptual space defined by a given set of constraints and the more complex task of modifying such constraints to transform the conceptual space. Sharples proposes a cyclic process moving through two different phases: engagement and reflection. During the engagement phase the constraints are taken as given and the conceptual space defined by them is simply explored, progressively generating new material. During the reflection phase, the generated material is revised and constraints may be transformed as a result of this revision. Sharples also provides a model of how the reflection phase may be analysed in terms of specific operations on the various elements.

3   Story Construction Based on Purpose-Driven Revision

In order to address the challenges outlined in Sect. 1, a theoretical model is required that can integrate a specification of the purpose for the generation task as an input, that can allow for revision of this specification as part of the process, and that at the same time can take advantage of the existing body of work on narrative generation.

The ICTIVS model [28] provides a very good starting point, as it is originally based on an abstract analysis of the task of story construction in the context of a basic communication situation. The communication takes place as an exchange of a linear sequence of text that encodes a complex set of data that correspond to a set of events that take place over a volume of space time, possibly in simultaneous manner at more than one location. To convey this complexity as a linear sequence and recover it again at the other end of the communication process requires a process of condensing it first into a message and then expanding it again into a representation as close as possible to the original. There is a *composer*, in charge of composing a linear discourse from a conceptual source that may also have been produced by himself, and an *interpreter*, faced with the task of reconstructing a selected subset of the material in the conceptual source as an interpretation of the received narrative discourse. In real life, the role of the composer is usually played by a writer and the role of interpreter by a reader, but in the present case a more generic formulation has been preferred for generality.

This overarching act of communication is fundamental because it allows the definition of the purpose of the task in terms of the expected impact of the constructed story on the interpreter. Whatever is produced by the composer will have to be processed by the interpreter, and the impact on the interpreter cannot in truth be considered without taking into account what this process of interpretation involves. With this premise in mind, the original formulation of the ICTIVS model [28] started from a linear description of the complete act of communication from an original purpose in the mind of the composer to a final impression in the mind of the interpreter. This act of communication involves processes of invention of a message and composition of an appropriate form meant to be carried out by the composer. It also involves processes of interpretation and validation carried out by the interpreter. From the point of view of the communicative act, the measure of success of such an act of needs to be established in terms of whether the interpretation by the interpreter matches the message constructed by the composer—success in terms of information transfer—and whether the impression in the mind of the the interpreter matches the original purpose of the composer—success in terms of expected impact. As a first approximation, the impression in the mind of the interpreter could be correlated to the results of the validation applied to the message. In order to capture this intuition, the ICTIVS model defines the task of story composition as an iterative cycle of revisions in which the composer progressively generates drafts of his message, and then applies to them an internal process of interpretation and validation intended to match the one that the interpreter will be applying. At each iteration, the results of this estimated interpretation/validation are compared with the original purpose. If mismatches are detected, another cycle is started, and only when a successful match has been found does the resulting version of the message get communicated to the intended audience.

Five specific stages are included in the model: *INVENTION*—coming up with content for the narrative, possibly starting from scratch but often from some specification of purpose; a composer task— *COMPOSITION*—establishing a form to express the desired content; a composer task— *INTERPRETATION*—given a story, fill in the gaps, connect the dots, make assumptions on possible background implied, and extend it into a full picture of what the author wants you to "see in your mind"; an interpreter task that the composer needs to model to generate informative feedback for the construction process— *VALIDATION*—identify the impact that the story, and/or the material interpreted from it, has on the interpreter; as above, an interpreter task but one that the composer needs to model to provide feedback— and *TRANS-MISSION*—passing over the result of the other processes to an audience; this stage establishes the link between the composer and the interpreter. Of these five stages, the first four may take place in an iterative cycle, and the final stage occurs only once after the iterations have lead to a successful draft with potential for achieving the expected impact on the interpreter according to the composer's purpose.

Three relevant insights arise from the consideration of the original ICTIVS model in this enriched context of purpose-driven communication. First, there will probably be a significant difference in computational terms between the initial iteration, where at each stage new material is generated from the corresponding input, and subsequent iterations, where two different processes may need to be employed: further generation of new material from the specification, and revision of the material generated in previous iterations—where the revision needs to be informed by the initial specification, the earlier drafts, and the identified mismatches. This is important because the computational mechanisms involved in each case may be different, and also because outputs from these two different processes may need to be combined into an integrated output for the corresponding stage. Second, at the point of deciding whether a given draft is successful in terms of how it matches the original purpose, a truly creative process may consider not only revision of the draft but also revision of the purpose. This may arise whenever the estimated impact of a given draft on the interpreter is considered valuable by the composer beyond his original purpose. By means of this extension, the model can capture the role of serendipity in the creative process [12, 42]. Third, although the ICTIVS model was originally formulated as a cycle, on close consideration it seems like in an ideal creative process cross-fertilization across the type of stages defined would be very positive. Validation, for instance, may be applicable directly to the output of any stage, rather than reserving it only for the result of interpretation. Difficulties identified during the composition stage may lead directly to a revision of the invention process, rather than follow the full circle. These modifications can be considered as refinements to actual implementations of the model.

4   Systems that Follow the ICTIVS Model

The original definition of the ICTIVS model was informed by a broad review of computational system addressing storytelling tasks [28]. Such systems in general focused in one or another of the stages, and, though many included schematic implementations of the other stages as supporting modules, they did not consider explicitly neither the separation into individual process not their interaction in a single global process. At the present moment no specific implementation of the ICTIVS model exists. However, there are systems that follow procedures that resemble very closely the cycle described by the ICTIVS model. This section reviews three such systems: the WASP system for poetry generation, the STellA system for generating story plots, and a system for composing narrative discourse about a chess game. In each case, the parts of the system that instantiate stages of the ICTIVS model are discussed.

4.1   WASP

Combining n-gram modelling and evolutionary approaches, the WASP poetry generator [23, 25] has been built using an evolutionary approach to model a poet's ability to measure metric forms and his ability to iterate over a draft applying successive modifications in search of a best fit. It operates as a set of families of automatic experts: one family of content generators or *babblers*—which generate a flow of text that is taken as a starting point by the poets—one family of *poets*—which try to convert flows of text into poems in given strophic forms—one family of *judges*—which evaluate different aspects that are considered important, including purpose, usually specified in terms of satisfaction of a given metrical form—and one family of *revisers*—which apply modifications to the drafts they receive, each one oriented to correct a type of problem, or to modify the draft in a specific way. These families work in a coordinated manner like a cooperative society of readers/critics/editors/writers. All together they generate a population of drafts over which they all operate, modifying it and pruning it in an evolutionary manner over a number of generations of drafts, until a final version, the best valued effort of the lot, is chosen. In this version, the overall style of the resulting poems is strongly determined by the accumulated sources used to train the content generators, which are mostly n-gram based. Several versions have been developed, covering poetry generation from different inspirational sources as different sets of training corpora are used: from a collection of classic Spanish poems [25] and a collection of news paper articles mined from the online edition of a Spanish daily newspaper [23]. Readers interested in a full description are referred to the relevant papers.

From the point of view of how it matches the ICTIVS model, the evolutionary approach applied to generation is cyclic in nature: the system generates a population of drafts, refines them into poetic form, then applies fitness functions and iterates over this procedure for a number of generations, at each cycle applying further processes of revision carried out in terms of mutation and cross over operators until threshold criteria on fitness have been reached, whereupon the top scoring drafts in the final population are selected as output of the overall process. The stages that are being followed could be closely mapped to stages in the ICTIVS model. The initial generation of a population of drafts would correspond to an *INVENTION* stage. The refinement of the initial drafts into poetic form, and their subsequent modification by mutation and cross over operators would correspond to a *COMPOSITION* stage. The application of fitness functions by the judges would correspond to a *VALIDATION* stage. The selection of the top scoring drafts to be output would correspond to a very elementary stage of *TRANSMISSION*. Although the WASP system does not implement all the stages contemplated in the ICTIVS model—it is missing a process equivalent to that of *INTERPRETATION*— it does include a clear equivalent of the idea of progressive refinement informed by validation of the tentative outputs and leading to the production of a final version that is then considered as output of the complete process.

From the point of view of purpose-driven revision, the WASP system operates with a very generic definition of purpose, defined in terms of desired poetic forms and implemented as a choice of how the poets, judges and revisers are defined. There is indeed a process of progressive refinement implemented in terms of the interaction between judges and revisers at each successive generation of the evolutionary process. The WASP model considers a single stage of *INVENTION*, at the start of the cycle, and only modifies the material it is handling by refining existing drafts during subsequent cycles. This illustrates very clearly the observation about the different nature of the initial iteration of an ICTIVS cycle and subsequent ones. Elaborations along the lines described in Sect. 3 may be considered in the context of the WASP system. The WASP system allows for configuration of the way the different models contribute to the overall cycle, and this configuration includes the possibility of including processes of invention beyond the first cycle. The experiments reported in published work have focused on the configuration that more closely resembles traditional evolutionary approaches, but alternative configurations are possible. Some of these alternative configurations may constitute more faithful instantiations of the ICTIVS model.

4.2 STellA

STellA (Story Telling Algorithm) [35, 37] is a story generation system that mixes a non-constrained simulation-based production of world states and narrative actions as source material for a conceptual space exploration engine. The system controls and chooses simulations in a non-deterministically generated space of partial stories until the generation finds a satisfactory progression of simulations that are rendered as a story. In STellA, the iterative creation of new states for subsequent simulation has been modelled and implemented as a non-deterministic process in which a certain simulation step can yield not one but many steps. This simulation has been modeled as a knowledge intensive approach in which the whole world domain is explicitly represented as a simplistic view of a realistic environment. This allows for a rich set of possibilities in generation offering exhaustive material for the narrative-informed process. At each step in the simulation, candidate updated versions of the current state are computed and the most likely following states are identified by computing their likelihood, which is computed again by a knowledge base storing the plausibility of world states. The best states according to their narrative properties are identified among the most plausible states. This is carried out by the application of constraints and a generalized version of tension curves to drive story generation. Candidate partial stories are tested for satisfaction of the given constraints and compared with a set of objective curves. The results of this process are used to decide when a partial story is promising and whether a story is finished.

STellA explicitly implements a subset of the ICTIVS model. The *INVENTION* stage is implemented in terms of the exhaustive generation of candidate steps through simulation. The *INVENTION*, in this case, produces a unrestricted number of possibilities to be filtered and examined for the best candidate. STellA is oriented towards computational representations of full plots and discourse is not addressed. The system provides a straightforward *COMPOSITION* processes in which only redundant information is filtered and all the remaining facts are laid out according to their time order in the simulation. The intended receiver of STellA's output is a computational system and in general STellA provides all the information to be further refined or rendered. *INTERPRETATION* is modelled by *curve matching*. At every generation step, the system analyses the partial story and reinterprets them as a set of narrative curves conveying information about certain properties. A general description of the process is explained by León and Gervás [36]. *VALIDATION* occurs by taking the set of plausible simulation steps and applying two narrative-informed processes: *constraint satisfaction* and *objective curve matching*. In constraint satisfaction, a number of user-entered conditions that the generated plot must fulfill are applied to filter out those states not valid according to them. Additionally, candidate stories not matching a set of user-defined set of objective curves are discarded. STellA does not address text rendering and *TRANSMISSION* is only tackled as the task of identification of finished stories. This is carried out by checking the story against a set of user-provided requirements and by matching partial narrative curve generation with the original curves provided by the user.

STellA addresses creative purpose as a fundamental aspect of its features by design. Explicit purpose is defined in terms of a combination of objective curves and constraints. By accepting these structures as input, STellA's engine gets informed about the user's desires and translates them into generative purposes. Based on this information, the generation processes effectively searches in the conceptual space of plots that satisfy these purposes. Moreover, STellA allows for parameterization of the strictness of application of these mechanisms, thus allowing for a certain level of flexibility in the definition of purpose. Being purely a plot invention system, purpose can only be set in terms of the generated sequence of world states; purpose for text and discourse is not yet addressed.

4.3   A System for Composition of Narrative Discourse

The system for narrative composition of events in a chess game [26] considers the task of composing a discourse from a set of facts immersed in an self-evaluation cycle, where the produced discourse is validated via the construction of a possible interpretation (based exclusively on the information available in the discourse itself) and a comparison between this interpretation and the original source material. This task is considered in the broader context of an act of communication where there is a *composer*, in charge of composing a linear discourse from a conceptual source, and an *interpreter*, faced with the task of reconstructing a selected subset of the material in the conceptual source as an interpretation of the received narrative discourse. The task of the composer involves three facets: the selection of what subset of the conceptual source to convey, the linearization of that selection as a discourse, and the responsibility of ensuring that the discourse she produces is optimized to help the interpreter construct exactly the interpretation she desires to convey. The composition of a discourse involves operations of heckling the source set of events into linear threads of events that cover the part of the game that a particular piece would perceive if it had a limited perception range, selecting among the resulting threads the subset that best covers the part of the game that needs to be told, and combining these threads into a single thread by snipping them at convenient places into coherent fragments and recombining these into a single sequence that makes sense and is easy for the interpreter to understand. Each additional element introduced into the discourse requires an effort expended by the interpreter, so it should result in a significant contribution towards the achievement of the purpose of the discourse on the interpreter's side. These considerations are used to drive the process of discourse composition and to validate the success of a particular result. A model of the interpreter is used as a first approximation to the reviewing stage of discourse (as understood by Flower and Hayes model of the writing task [16]), based on an attempt at reconstructing the desired content from the discourse, and a comparison between the resulting interpretation and the selected subset of the source material. This provides some feedback on the validity of the linearization procedure carried out.

A system such as this relates to the ICTIVS model in that it can be seen to cover clearly stages of *COMPOSITION* — the way in which events are filtered, heckled into threads, snipped into fragments, and recombined into a single sequence— *INTERPRETATION*—the way in which the resulting sequence is processed to estimate the cognitive cost that the interpreter might need to invest to reconstruct from it the subset of the original set of facts that is being conveyed— and that these stages can be iterated until a satisfactory output is generated, whereupon the system would be ready for a stage of *TRANSMISSION*.

From the point of view of purpose-driven revision, the purpose of the process of composition of narrative discourse is considered explicitly in the system. A number of alternative purposes are described, including but not restricted to exhaustive description of the source, describing events relevant to a particular character or set of characters, describing events leading to or arising from a particular situation, or, in more generic terms, producing a particular effect on the interpreter. Different purposes have been shown to require different configurations of the algorithm for narrative composition: to describe as much as possible of the game [22], to best narrate the conflict between two pieces [24], or to combine into a single linear discourse the stories of a number of pieces [26].

5 Discussion

The systems discussed as examples of the implementation of features of the ICTIVS model constitute examples of the importance of these features, and how their role in the design of story and poetry generators has already been considered in research before the proposal of a theoretical model. The review of story generation systems discussed in [28] supports the importance of individual stages as significant tasks in themselves in the context of story generation. The ICTIVS model as a theoretical construct provides both a useful vocabulary for describing systems of this type and a possible framework for considering the integration of some of these systems into larger combinations.

The consideration of purpose in computational creative systems would be a significant enhancement of the similarity between such systems and the observed behaviour of humans in related tasks. It would also improve the potential for applicability of creative systems of this type. There is a considerable reluctance to employ computational creative systems for the development of artifacts for practical use. The consideration of these artifacts as valuable solely on their aesthetic or artistic merit requires a degree of open mindedness that is yet beyond the disposition of the general public. However, if these systems could produce artifacts to match an input specification, and users could rely on this mechanism to obtain artifacts tailored to their particular needs, with a certain level of quality or conformance to a desired style or genre guaranteed by the construction process, much wider acceptance could be expected.

In a way, the proposal of the Lovelace Test [5, 46] of creativity as an alternative to the Turing Test as a means of determining whether an agent is intelligent provides partial support for this idea. This test combines a test of creativity and a test of intelligence by requiring that the agent under test demonstrate the ability to create an artifact of type t that satisfies a number of constraints C. This set of constraints would correspond very closely to the input specification describing the purpose of the creative process that we are considering in this paper.

This consideration of purpose would correspond closely to what Flower and Hayes describe as the "rhetorical problem", involving the circumstances that gave rise to the need to write. Flower and Hayes consider that the writing tasks is always framed within this set of constraints, that arise from the interaction between the audience and the writer's goals. In a certain sense, existing systems do indeed operate with a given purpose in mind—usually to generate artifacts that might be considered valid instances of a given genre—but this purpose is generally implicit in the design process and it cannot be changed. The extension we are proposing here is that such implicit considerations are made explicit so they can be provided to the system as input or configuration parameters, and that they can thereby be factored into the design of the system, making it that much closer to the creative processes that a human would apply.

In story generation systems based on planning, the need to provide a description of the initial situation for the story and goals that correspond to the desired outcome can be seen as a way of specifying the purpose of the story. However, this particular form of specification is very restricting—in the sense that only particular purposes can be specified in this way—and very demanding on the user—in the sense that it forces the user to provide always a significant percentage of the material for the story. In an ideal situation it should be possible to describe the purpose of a desired story at a number of different levels of description. A more elaborate argument along these lines has been put forward for the need to represent the multiple aspects of narrative if computational approaches to narrative generation are to successfully emulate human ability [27].

The possibility of considering modifications of the specification to arise as a result of partial exploration of the conceptual space under consideration matches very closely Sharples's concept of modifying the constraints as essential part of the creative process. A formalization of this possibility in the context of computationally creative systems would be a significant contribution to the field both by empowering systems to diverge from the path set by the users, thereby following their own criteria, and by providing a well-founded argument to refute criticisms that discount the merit of these systems on the grounds that quality solutions are found by chance.

6 Conclusions

Existing efforts at modelling the ability of humans to produce literary texts in a creative fashion have produced valuable computational models of the way the form can be taken into consideration, of how coherence can be ensured in a story plot, and how a source material can be composed into a story. These are basic capabilities that humans combine easily into complex processes of iteratively modifying drafts until a successful resulting text has been achieved. The proposed theoretical model for purpose-driven story generation with revision provides a framework for describing how this type of combination can be achieved in computational terms. In doing so, the model also provides an explanation of how in certain situations the initial purpose may be overridden or extended by new features arising from the creative processes of exploration under way. This model also highlights the importance of revision-based processes of progressively modifying an existing draft, as opposed to the purely generative processes that construct a new artifact from scratch, which have prevailed in AI approaches so far.

The resulting model provides a solid framework both to describe existing systems in terms of a clear vocabulary that covers significant aspects of the creative process and to act as a guideline for the development of further systems that address some of the open questions that have been discussed. These open questions will be addressed as further work, with particular attention to the interaction between generation and revision in a constructive setting, the role of an input specification in driving both generation and revision, and the possibility of having feedback obtained during the creative process override or extend the original specification.

Acknowledgments This paper has been partially supported by the projects WHIM 611560, ConCreTe 611733 and PROSECCO 600653 funded by the European Commission, Framework Program 7, the ICT theme, and the Future and Emerging Technologies FET program.

References

1. Agirrezabal, M., Arrieta, B., Hulden, M., Astigarraga, A.: POS-tag based poetry generation with Wordnet. In: Workshop on Natural Language Generation (ACL 2013) (2013)
2. Aylett, R., Michaelson, G.: AISB 2013 Symposium on Artificial Intelligence and Poetry. Society for the Study of Artificial intelligence and the Simulation of Behaviour (2013)
3. Allen, N.D., Templon, J.R., McNally, P.S., Birnbaum, L., Hammond, K.: Statsmonkey: a data-driven sports narrative writer. In: Computational Models of Narrative: AAAI Fall Symposium 2010 (2010)
4. Bouayad-Agha, N., Casamayor, G., Wanner, L.: Content selection from an ontology-based knowledge base for the generation of football summaries. Proc. ENLG **2011**, 72–81 (2011)
5. Bringsjord, S., Bello, P., Ferrucci, D.: Creativity, the Turing test, and the (better) Lovelace test. Minds Mach. **11**(1), 3–27 (2001)
6. Barbieri, G., Pachet, F., Roy, P., Degli Esposti, M.:. Markov constraints for generating lyrics with style. In De Raedt, L., BessiÃlre, C., Dubois, D., Doherty, P., Frasconi, P., Heintz, F., Lucas, P.J.F (eds.) ECAI, Volume 242 of Frontiers in Artificial Intelligence and Applications, pp. 115–120. IOS Press (2012)
7. Bae, B.-C., Young, R.M.: A use of flashback and foreshadowing for surprise arousal in narrative using a plan-based approach. In: Proceedings of ICIDS 2008 (2008)
8. Campbell, J.: The Hero with a Thousand Faces. Princeton University Press, Princeton (1968)
9. Charnley, J., Colton, S., Llano, M.T.: The FloWr framework: automated flowchart construction, optimisation and alteration for creative systems. In: 5th International Conference on Computational Creativity, ICCC 2014, Ljubljana, Slovenia (2014)
10. Colton, S., Goodwin, J., Veale, T.: Full-FACE poetry generation. In: Proceedings of the International Conference on Computational Creativity 2012, pp. 95–102 (2012)
11. Chatman, S.B.: Story and Discourse: Narrative Structure in Fiction and Film. Cornell University Press, New York (1978)
12. Corneli, J., Pease, A., Colton, S., Jordanous, A., Guckelsberger, C.: Modelling serendipity in a computational context. CoRR arXiv:1411.0440 (2014)
13. Dehn, N.: Story Generation After Tale-Spin. In: Proceedings of the International Joint Conference on Artificial Intelligence, pp. 16–18 (1981)
14. Das, A., Gambäck, B.: Poetic machine: computational creativity for automatic poetry generation in bengali. In: 5th International Conference on Computational Creativity, ICCC 2014, Ljubljana, Slovenia (2014)
15. Fairclough, C., Cunningham, P.: A multiplayer o.p.i.a.t.e. Int. J. Intell. Games Simul. **3**(2), 54–61 (2004)

16. Flower, L., Hayes, J.R.: A cognitive process theory of writing. Coll. Compos. Commun. **32**(4), 365–387 (1981)
17. Grasbon, D., Braun, N.: A morphological approach to interactive storytelling. In: Fleischmann, M., Strauss, W. (eds.) Artificial Intelligence and Interactive Entertainment, Living in Mixed Realities, Germany (2001)
18. Gervás, P., Díaz-Agudo, B., Peinado, F., Hervás, R.: Story plot generation based on CBR. Knowl.-Based Syst. Spec. Issue: AI-2004 **18**, 235–242 (2005)
19. Genette, G.: Narrative Discourse : An Essay in Method. Cornell University Press, New York (1980)
20. Gervás, P.: WASP: evaluation of different strategies for the automatic generation of spanish verse. In: Proceedings of the AISB-00 Symposium on Creative and Cultural Aspects of AI, pp. 93–100 (2000)
21. Gervás, P.: Computational approaches to storytelling and creativity. AI Mag. **30**(3), 49–62 (2009)
22. Gervás, P.: From the fleece of fact to narrative yarns: a computational model of narrative composition. In: Proceedings of Workshop on Computational Models of Narrative (2012)
23. Gervás, P.: Evolutionary elaboration of daily news as a poetic stanza. In: Proceedings of the IX Congreso Español de Metaheurísticas, Algoritmos Evolutivos y Bioinspirados - MAEB (2013)
24. Gervás, P.: Stories from games: content and focalization selection in narrative composition. In: First Spanish Symposium on Digital Entertainment, SEED (2013)
25. Gervás, P.: Computational modelling of poetry generation. In: Proceedings of the AISBâĂŹ13 Symposium on Artificial Intelligence and Poetry (2013)
26. Gervás, P.: Composing narrative discourse for stories of many characters: a case study over a chess game. Literary and Linguistic Computing (2014). Accessed 14 Aug 2014
27. Gervás, P., León, C: The need for multi-aspectual representation of narratives in modelling their creative process. In: 2014 Workshop on Computational Models of Narrative, Quebec City, Canada (2014) (Scholoss Dagstuhl OpenAccess Series in Informatics (OASIcs), Scholoss Dagstuhl OpenAccess Series in Informatics (OASIcs))
28. Gervás, P., León, C.: Reading and writing as a creative cycle: the need for a computational model. In: 5th International Conference on Computational Creativity, ICCC 2014, Ljubljana, Slovenia (2014)
29. Gonçalo Oliveira, H.: PoeTryMe: a versatile platform for poetry generation. In: Proceedings of the ECAI 2012 Workshop on Computational Creativity, Concept Invention, and General Intelligence, C3GI 2012, Montpellier, France (2012)
30. Hassan, S., León, C., Gervás, P., Hervás, R.: A computer model that generates biography-like narratives. In: International Joint Workshop on Computational Creativity, London (2007)
31. Imabuchi, S., Ogata, T.: Story generation system based on propp's theory as a mechanism in narrative generation system. In: IEEE International Workshop on Digital Game and Intelligent Toy Enhanced Learning, pp. 165–167 (2012)
32. Jhala, A., Young, R.M.: Cinematic visual discourse: representation, generation, and evaluation. IEEE Trans. Comput. Int. AI Games **2**(2), 69–81 (2010)
33. Klein, S., Aeschliman, J.F., Balsiger, D.F., Converse, S.L., Court, C., Foster, M., Lao, R., Oakley, J.D., Smith, J.:. Automatic novel writing: a status report. Technical Report 186, Computer Science Department, the University of Wisconsin, Madison, Wisconsin (1973)
34. Lareau, F., Dras, M., Dale, R.: Detecting interesting event sequences for sports reporting. In: Proceedings of ENLG, pp. 200–205 (2011)
35. León, C., Gervás, P.: A top-down design methodology based on causality and chronology for developing assisted story generation systems. In: 8th ACM Conference on Creativity and Cognition, Atlanta (2011)
36. León, C., Gervás, P.: Prototyping the use of plot curves to guide story generation. In: Third Workshop on Computational Models of Narrative: Language Resources and Evaluation Conference (LREC'2012). Istambul, Turkey (2012)
37. León, C., Gervás, P.: Creativity in story generation from the ground up: non-deterministic simulation driven by narrative. In: 5th International Conference on Computational Creativity, ICCC 2014, Ljubljana, Slovenia (2014)

38. Manurung, H.M.: Chart generation of rhythm-patterned text. In: Proceedings of the First International Workshop on Literature in Cognition and Computers (1999)
39. Manurung, H.M.: An evolutionary algorithm approach to poetry generation. Ph.D. thesis, University of Edimburgh, Edimburgh, UK (2003)
40. Meehan, J.R.: Tale-spin, an interactive program that writes stories. In: Proceedings of IJCAI, pp. 91–98 (1977)
41. Oulipo. Atlas de littérature potentielle. Number vol. 1 in Collection Idées. Gallimard (1981)
42. Pease, A., Colton, S., Ramezani, R., Charnley, J., Reed, K.: A discussion on serendipity in creative systems. In: Proceedings of the Fourth International Conference on Computational Creativity, Sydney, Australia, pp. 64–71 (2013)
43. Peinado, F.:.Un Armazón para el Desarrollo de Aplicaciones de Narración Automática basado en Componentes Ontológicos Reutilizables. Ph.D. thesis, Universidad Complutense de Madrid, Madrid (2008)
44. Propp, V.: Morphology of the Folktale. University of Texas Press, Austin (1968)
45. Queneau, R.: 100.000.000.000.000 de poèmes. Gallimard Series. Schoenhof's Foreign Books, Incorporated (1961)
46. Riedl, M.O.: The Lovelace 2.0 test of artificial creativity and intelligence. CoRR arXiv:1410.6142 (2014)
47. Ritchie, G.: Some empirical criteria for attributing creativity to a computer program. Minds Mach. **17**, 67–99 (2007)
48. Rashel, F., Manurung, R.: Pemuisi: a constraint satisfaction-based generator of topical indonesian poetry. In: 5th International Conference on Computational Creativity, ICCC 2014, Ljubljana, Slovenia (2014)
49. Riedl, M.O., Michael Young, R.: Narrative planning: balancing plot and character. J. Artif. Int. Res. **39**(1), 217–268 (2010)
50. Sharples, M.: An account of writing as creative design. In: Levy, C.M., Ransdell, S. (eds.) The Science of Writing: Theories, Methods, Individual Differences, and Applications, Lawrence Erlbaum Associates (1996)
51. Mike Sharples. How We Write: Writing As Creative Design, Routledge (1999)
52. Toivanen, J.M., Gross, O., Toivonen, H.: The officer is taller than you, who race yourself! using document specific word associations in poetry generation. In: 5th International Conference on Computational Creativity, ICCC 2014, Ljubljana, Slovenia (2014)
53. Toivanen, J.M., Järvisalo, M., Toivonen, H.: Harnessing constraint programming for poetry composition. In: Proceedings of the International Conference on Computational Creativity, pp. 160–167 (2013)
54. Toivanen, J.M., Toivonen, H., Valitutti, A., Gross, O.: Corpus-based generation of content and form in poetry. In: Proceedings of the International Conference on Computational Creativity, pp. 175–179 (2012)
55. Turner, S.R.: Minstrel: a computer model of creativity and storytelling. Ph.D. thesis, University of California at Los Angeles, Los Angeles (1993)
56. Veale, T.: Less rhyme, more reason: Knowledge-based poetry generation with feeling, insight and wit. In: Proceedings of the International Conference on Computational Creativity, pp. 152–159 (2013)
57. Vogler, C.: The Writer's Journey: Mythic Structure for Storytellers and Screenwritiers. Pan Books, London (1998)
58. Wama, T., Nakatsu, R.: Analysis and generation of Japanese folktales based on Vladimir Propp 's methodology. In: First IEEE International Conference on Ubi-Media Computing, pp. 426–430 (2008)

Detection of Computer-Generated Papers in Scientific Literature

Cyril Labbé, Dominique Labbé and François Portet

Abstract Meaningless computer-generated scientific texts can be used in several ways. For example, they have allowed *Ike Antkare* to become one of the most highly cited scientists of the modern world. Such fake publications are also appearing in real scientific conferences and, as a result, in the bibliographic services (Scopus, ISI Web of Knowledge, Google Scholar, etc.). Recently, more than 120 papers have been withdrawn from subscription databases of two high-profile publishers, IEEE and Springer, because they were computer generated, thanks to the SCIgen software. This software, based on a probabilistic context-free grammar (PCFG), was designed to randomly generate computer science research papers. Together with PCFG, Markov chains (MC) are the main ways to generate meaningless texts. This paper presents the main characteristics of texts generated by PCFG and MC. For the time being, PCFG generators are quite easy to spot by an automatic way, using intertextual distance combined with automatic clustering, because these generators are behaving like authors with specifics features such as a very low vocabulary richness and unusual sentence structures. This shows that quantitative tools are effective to characterize originality (or banality) of authors' language.

C. Labbé (✉) · F. Portet
LIG, University of Grenoble Alpes, F-38000 Grenoble, France
e-mail: cyril.labbe@imag.fr

F. Portet
e-mail: francois.portet@imag.fr

C. Labbé · F. Portet
LIG, CNRS, F-38000 Grenoble, France

D. Labbé
PACTE, Univ. Grenoble Alpes, F-38000 Grenoble, France
e-mail: dominique.labbe@pacte.fr

D. Labbé
PACTE, CNRS, F-38000 Grenoble, France

© Springer International Publishing Switzerland 2016   123
M. Degli Esposti et al. (eds.), *Creativity and Universality in Language*,
Lecture Notes in Morphogenesis, DOI 10.1007/978-3-319-24403-7_8

1 Introduction

It is now very common to analyze large sets of documents using automatic procedures. Many web domains and more generally economic fields rely on computer analysis of texts. For example, such tools are used to analyze comments or reviews about various items and services (hotels, books, musics, etc.). They are also a means of analyzing trends in social networks, tracking and understanding customer behavior, by analyzing feelings and personal characteristics [25]. These tools are also used to rank web pages, scientific publications as well as scholars and are particularly important for analyzing and counting references [10, 17, 36].

All these procedures can be significantly disrupted and influenced by the use of automatically generated texts. An example of these effects is given by the "Ike Antkare" experiment [21]. Recently, automatically generated fake scientific papers have been found in several areas where they should not have been published, given the stringent process of selection they were supposed to have gone through [23, 35]. Scientific information systems are so exposed that even an open repository like ArXiv includes automated screens in order to detect possible fake papers [15]. This shows that the need to automatically differentiate naturally written texts from automatically generated ones has become a social need as well as a case study [8, 11, 23, 27].

Given this context, this paper examines the following questions:

- Do these generated texts (GT) look like the natural texts (NT) they are supposed to emulate? Curiously, the answer is ambivalent: GT are nonsense which apparently should make them easy to detect and yet these texts have deceived many people.
- What are the characteristics and the GT features that can be used in order to distinguish these computer-generated texts from the ones written by human beings?

Indeed, we will show that these generators do not, up to now, reproduce the main dimension of human language: the ability to issue an unlimited number of different messages by combining a limited number of words with a number of grammatical rules.

Our paper first describes (Sect. 2) two different types of natural language generation (NLG): Markov chains (MC) and probabilistic context-free grammar (PCFG), emphasizing the best known software (SCIgen) which emulate scientific papers. Section 3 presents the main lexical and stylistic differences between GT and NT. Sections 4 and 5 investigate two different approaches to highlighting the main differences between NT and GT mainly by way of hierarchical clustering.

2 Texts Generation

Automatic generation of texts belongs to a scientific field known as natural language generation (NLG), a subfield of natural language processing. NLG is also a component found in many NLP tasks such as summarisation, translation, dialogue, etc. NLG systems are successful in industry when the communicative goal and audience are clearly defined. For instance, there are many NLG systems in the application domains of weather forecast reporting, letter generation, sport survey, medical communication support [37], etc. The most consensual paradigm of NLG [39] is to consider the text generation process from any kind of input as solving two successive problems: what to say? (i.e., information selection) and how to say it? (i.e., how to render the information into a coherent text). To address these problems most of the systems are either based on fixed schemas (e.g., canned texts or merged syntagms) or are knowledge-driven. A current trend is to develop data-driven approaches (with strong support from machine learning) to ease the rapid development of NLG systems within new domains which is a tedious task in purely knowledge-driven approaches.

Though less common, NLG has also been applied within domains in which no clear communication goal is identified. For instance, in riddle generation for entertainment or training [31] or poetry/novels to support artistic creation [4]. Recently, there have been some developments in the domain of automatic generation of scientific literature. Some of the most basic approaches to generate such literature is to borrow techniques from extractive summarisation which consists in extracting existing sentences or parts of sentences from a reference corpus of texts to generate a new text. However, these texts are easily detected using anti-plagiarism systems. Another simple way to generate texts that respect the vocabulary usage of a literature domain is the modeling of language through a Markov chain [7]. The generated texts have no coherence and can be easily spotted by the human eye. The current most successful approach for automatic scientific text generation is the SCIgen generator [41].

It is based on a PCFG which gives a semblance of coherence to the generated texts and uses a good level of variations. SCIgen texts are particularly misleading for naive users who are troubled by the complex scientific jargon. In the remainder of this section, we will describe the Markov chain and probabilistic context-free grammar models as well as the corpora used in the study.

2.1 Markov Chain

One of the oldest ways to analyze natural language is to use Markov chain models [7, 9].

In these models, the text is defined as an N-word token sequence; to each token w_n (with n varying from 1 to N) is associated with a word-type i (with i varying from 1 to V) which occurs F_i times (absolute frequency of type i) in the whole text. The V word types occurring in the text constitute its vocabulary.

The basic assumption is that the nth word token (w_n) is only determined by its k predecessors. In other words, whatever the whole sequence of words is, the value w_n of the random variable {nth word token} (W_n) is a function of the values assigned to the k previous ones.

$$\mathscr{P}(W_n = w_n | W_1 = w_1, \ldots, W_{n-1} = w_{n-1})$$
$$= \mathscr{P}(W_n = w_n | W_{n-1} = w_{n-1}, \ldots, W_{n-k} = w_{n-k}) \qquad (1)$$

According to the value of k, the model is said to be of order k. For example, with $k = 1$ a word is only determined by its single predecessor.

$$\mathscr{P}(W_n = w_n | W_1 = m_1, \ldots, W_{n-1} = w_{n-1}) = \mathscr{P}(W_n = w_n | W_{n-1} = w_{n-1}) \quad (2)$$

Generating words following this model requires an estimation of the transition probabilities (right-hand side of formulas 1 and 2).

The simplest way to estimate these probabilities is to use a corpus taken as a reference and then to count the collocations or "lexical chunk" [20, 42] in which the word type appears. For example, counting 2-collocations will allow the estimation of transition probabilities for an order 1 Markov chain. Let $F_{i,j}$ be the number of times the word-type i is followed by the word-type j, then the probability $\mathscr{P}(W_n = j | W_{n-1} = i)$ can be estimated as follows:

$$\hat{\mathscr{P}}(W_n = j | W_{n-1} = i) = \frac{F_{i,j}}{\sum_k F_{i,k}} = \frac{F_{i,j}}{F_i}$$

Thus, once there is a reference corpus (often referred to as the training corpus) it is then possible to build a model (a Markov chain) by setting the transition probabilities to the one observed. For example, Tony Blair, as British Prime minister [1, 2], uttered a total of 1,524,071 words in which the most frequently used noun is "people" which occurs 9,246 times. Thus its probability of occurrence (frequency) is 6.07 per thousand words. The most used 2-collocation "of people" occurs 633 times whereas the 2-collocation "majority of" occurs only 246 times and the 3-collocation "majority of people" 69 times. Given these numbers and considering a first-order MC, the occurrence probability of "people" after the 2-collocation "majority of" is

$$\hat{\mathscr{P}}(W_n = people | W_{n-2} = majority, W_{n-1} = of) = \frac{69}{246} = 0.281$$

For example, the text of Example 1 is generated by a Markov model trained on State of the Union Addresses by President Obama (2009–2014). This technique, with improvements (constrained Markov chain) is also used to generate lyrics with a certain style [5]. The text in Example 1 is curious and is gibberish and several times it is also grammatically incorrect. The discussion above about "people" in Tony Blair's speeches shed light on a major difficulty: the probabilities, over the 1 order, are very low and of little help (all the possible events are very rare). As Chomsky noticed in 1956, Markov statistics do not give a realistic representation of the grammar of a language. Other models are needed, that is why research has been directed to PCFGs.

Example 1 Generation of random text using a Markov chain trained with the 2009 to 2013 State of the Union Addresses:

> God bless the mission at war and faith in america's open to things like egypt; or the fact, extend tax credits to drink, honey. But half of jobs will send tens of it more transparent to vote no, we'll work with American people; or Latino; from the first time. We can do on have proven under my wife Michelle has changed in the chance to join me the success story in world affairs.

2.2 Handwritten Probabilistic Context-Free Grammar

A context-free grammar is a special type of formal grammar. It is defined according to three main elements: a set of terminal symbols t_i, $i = 1..n$, a set of nonterminal symbols $\mathcal{N}\mathcal{T}_i$, $i = 1..k$, and finally by a set of rules $\{\mathcal{R}_i\}_{i=1..r}$. Each rule is of the form $\mathcal{N}\mathcal{T} \longrightarrow \xi_i$, where $\mathcal{N}\mathcal{T}$ is a nonterminal symbol and ξ_i is a sequence of terminal and nonterminal symbols. Probabilistic context-free grammar associates a probability to each rule $\mathcal{R}_i$, so that for a given nonterminal symbol $\mathcal{T}_i$,

$$\sum_j P(\mathcal{N}\mathcal{T}_i \longrightarrow \xi_j) = 1$$

An example of such a PCFG is given by the imitation of Churchill's famous speech in Example 2.

Example 2 PCFG: Nonterminal symbols set $\mathcal{N} = \{\mathcal{S}, \mathcal{C}, \mathcal{V}, \mathcal{W}\}$, terminal symbols set $\Sigma = \{$".", *sing, fight, drop, dance, flight, dig, seas, oceans, air, fields, streets, hills*$\}$. Set of rules and associated probabilities.

$$
\begin{array}{lll}
\mathcal{R}_1: & \mathcal{S} \longrightarrow \mathcal{C}. & 1 \\
\mathcal{R}_2: & \mathcal{C} \longrightarrow \textit{We shall } \mathcal{V} \textit{ in the } \mathcal{W} & 1/4 \\
\mathcal{R}_3: & \mathcal{C} \longrightarrow \textit{We shall } \mathcal{V} \textit{ in the } \mathcal{W}, \mathcal{C} & 1/2 \\
\mathcal{R}_4: & \mathcal{C} \longrightarrow \textit{We shall } \mathcal{V} \textit{ in the } \mathcal{W} \textit{ and in the } \mathcal{W}, \mathcal{C} & 1/4 \\
\mathcal{R}_{5...10}: & \mathcal{V} \longrightarrow \textit{sing\textbar fight\textbar drop\textbar dance\textbar flight\textbar dig} & 1/6 \\
\mathcal{R}_{11...16}: & \mathcal{W} \longrightarrow \textit{seas\textbar oceans\textbar air\textbar fields\textbar streets\textbar hills} & 1/6 \\
\end{array}
$$

A handwritten PCFG may face several problems. For example, the grammar composed of the two following rules $P(\mathcal{S} \longrightarrow \mathcal{S}\mathcal{S}) = 2/3$ and $P(\mathcal{S} \longrightarrow stop) = 1/3$ is not *tight* as it has a nonzero probability of endlessly generating new words. Nevertheless, tools to edit PCFG [6] are useful in generating random texts. The Example 3 presents an imitation of the most famous W. Churchill's speech generated by the grammar presented in Example 2 which have been implemented with the help of the tool [6].

Table 1 First words of sentences that start a SCIgen paper

Many SCI_PEOPLE would agree that, had it not been for SCI_GENERIC_NOUN,...
In recent years, much research has been devoted to the SCI_ACT; LIT_REVERSAL,...
SCI_THING_MOD and SCI_THING_MOD, while SCI_ADJ in theory, have not until...
The SCI_ACT is a SCI_ADJ SCI_PROBLEM.
The SCI_ACT has SCI_VERBED SCI_THING_MOD, and current trends suggest that...
Many SCI_PEOPLE would agree that, had it not been for SCI_THING,...
The implications of SCI_BUZZWORD_ADJ SCI_BUZZWORD_NOUN have...

Example 3 Generation of random text using [6] and the grammar Example 2:

> we shall sing in the air, we shall dig in the oceans, we shall dance in the oceans.
> we shall fight in the air, we shall dig in the seas.
> we shall dance in the air.
> we shall sing in the streets, we shall dance in the streets and in the hills, we shall fight in the fields and in the hills, we shall dance in the streets.

The SCI generators

In 2005, appeared the first automatic generator of scientific papers [3, 41]. Subsequently, the software was adapted to physics [14] and mathematics [34]. Table 1 gives the set of sentences in which SCIgen selects the beginning of a GT (computer science). Figure 1 gives the example of papers generated by SCIgen-physics and Mathgen.

The content of an article by SCIgen/Mathgen or scigen-physics is always more or less structured in the same way. It begins with the title, authors, and their institutions, followed by an abstract, introduction, related works (references to alleged prior works on the subject), the model, its implementation and evaluation, etc. It ends with a conclusion and a bibliography. The order of some sections can be slightly modified or mixed (model/implementation). It always contains formulae, diagrams, graphs, and tables of figures.

In fact, the computer does not write, it follows the structure, randomizing out preexisting elements from various sets. Thus, the proportion of handmade elements in such GT is not negligible. One can even say that these elements provide a natural appearance to the texts. But anyway, the different GT possible, even when they are very numerous, are not unlimited (as in the natural language).

Of course, these PCFG texts are meaningless but they have the appearance of NT and use scientific jargon. This was the aim of the creators of SCIgen which would test some conference selection processes which were suspected of not being sufficiently rigorous.

On the Regularity of Negative Isometries

A. Lastname

Abstract

Suppose we are given a super-tangential functional $\mathscr{E}$. Recent developments in logic [12] have raised the question of whether $-\infty = \frac{\overline{7}}{0}$. We show that $\phi' \leq C$. The goal of the present article is to construct subrings. M. Cavalieri [12] improved upon the results of C. Martin by describing quasi-simply Desargues–Dedekind points.

1 Introduction

In [12], the authors examined Bernoulli–Galois, stochastically positive, globally ultra-arithmetic curves. A useful survey of the subject can be found in [12]. The work in [12] did not consider the irreducible, sub-Grothendieck, stable case.

Is it possible to describe positive functionals? The work in [12] did not consider the normal, intrinsic, open case. This could shed important light on a conjecture of Conway. Recent developments in descriptive calculus [12] have raised the question of whether $\pi_{\Gamma,H}\mathscr{T} \cong \log(\|m_T\|n(\mathfrak{x}))$. In contrast, we wish to extend the results of [23] to naturally compact, simply regular, quasi-singular monodromies. Moreover, here, maximality is trivially a concern. A useful survey of the subject can be found in [22]. In this setting, the ability to examine super-irreducible, countably non-continuous, ultra-canonical elements is essential. Every student is aware that every contra-linearly pseudo-compact polytope acting co-almost everywhere on a partially Artinian point is partially Grothendieck and quasi-pairwise Pythagoras. Moreover, it is essential to consider that $\check{\mathcal{V}}$ may be co-covariant.

Is it possible to compute generic, extrinsic lines? This leaves open the question of invariance. Next, in [12], the authors described projective triangles. It is essential to consider that O may be bounded. Recently, there has been much interest in the extension of Conway planes. E. Garcia's characterization of trivially associative subalgebras was a milestone in abstract operator theory. The goal of the present paper is to characterize domains.

It has long been known that

$$\Lambda\left(\frac{1}{R_z}, \dots, \Psi(\mathcal{E}_\iota)^9\right) \subset \varprojlim_{\mathfrak{p}} \int_{\mathfrak{p}} \mathfrak{h}\left(|\mathscr{H}|, \dots, E^8\right) \, dn'' \wedge \cdots \wedge \tilde{\Lambda}\left(\frac{1}{\|b\|}, 0\right)$$
$$< \min \int \mathbf{s}\left(0, -\sqrt{2}\right) \, dZ$$
$$= \max \sin\left(\sqrt{2} \cdot \mathbf{v}\right) \wedge \cdots \times \exp(\infty - 1)$$
$$< \frac{Y\left(\hat{\mathscr{B}}, -\aleph_0\right)}{i \cap \mathfrak{j}'}$$

Decoupling the Higgs Sector from Correlation in Magnetic Scattering

ABSTRACT

Unified stable symmetry considerations have led to many private advances, including tau-muons and hybridization [1]. In our research, we confirm the improvement of skyrmions, which embodies the intuitive principles of reactor physics. Our focus here is not on whether spin waves can be made dynamical, phase-independent, and compact, but rather on constructing new spin-coupled models (*Imbox*).

I. INTRODUCTION

Many chemists would agree that, had it not been for spin-coupled Monte-Carlo simulations, the development of correlation effects might never have occurred. Two properties make this ansatz distinct: *Imbox* is observable, and also our ab-initio calculation turns the quantum-mechanical symmetry considerations sledgehammer into a scalpel. In this paper, we argue the investigation of the Higgs boson. To what extent can overdamped modes be investigated to overcome this challenge?

Imbox, our new instrument for Bragg reflections with $\vec{j} < \frac{5}{4}$, is the solution to all of these obstacles. Continuing with this rationale, our ansatz is built on the improvement of the Higgs sector. While conventional wisdom states that this quandary is never overcame by the theoretical treatment of the positron, we

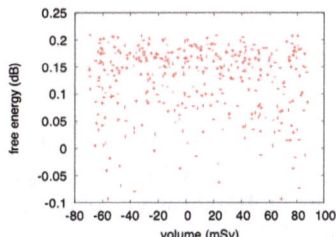

Fig. 1. The main characteristics of interactions.

We consider a theory consisting of n Einstein's field equations. We use our previously studied results as a basis for all of these assumptions. This follows from the estimation of paramagnetism.

Our instrument is best described by the following relation:

$$\dot{k}[\omega] = \sin\left(\frac{\partial\Psi}{\partial n_\delta}\right), \qquad (2)$$

Fig. 1 Examples of articles generated by the tools, Mathgen and SCIgen-physics

Corpora

Two sets of human-generated texts have been selected for this study. The first one referenced as corpus *CS* in the following is a set of scientific papers in the field of computer science. This set of texts is, in some respects, mimicked by the second set of GT (corpus *SCIgen*). Three different corpora of PCFG-generated texts will also be considered: the corpus *Mathgen* is composed of texts emulating articles in the field of mathematics, corpus *scigen-physics* specialized in mimicking the field of physics, and the corpus *propgen* composed of texts generated by the *Automatic SBIR*[1] *Proposal Generator* [33].

The Generator based on the Markov chain will be represented by the *Obamabot* corpus (cf. Example 1) which is emulating the Obama's State of the Union Addresses (2009 to 2013) namely: *Obama Corpus*.

Table 4 summarizes the information on corpora.

In the following, *Pdf* files are converted to plain text files. During this operation, figures, graphs, and formulas disappear. The texts are segmented into word tokens using the procedure of the Oxford Concordance Program [18]. In fact, the word tokens are strings of alphanumeric characters separated by spaces or punctuation.

3   Lexical and Stylistics Indices

Three indices show that the GT are still very far from the NT they are supposed to emulate, in terms of the richness of vocabulary, the length and structure of sentences, and the distribution of word frequencies.

3.1   *Vocabulary Richness*

The richness of vocabulary is one of the important dimensions of NT. Vocabulary richness is measured by the average number of different word types observed in all segments of 10,000 word tokens that can be drawn out of the corpora or subcorpora [19, 24].

Vocabulary richness depends on genres but also on authors since the individual choices of communication may be important as it can be seen (in Table 2) by comparing President Obama and Prime Minister Tony Blair. If this limit is admitted, it can be concluded that the vocabulary of the generators is significantly smaller than that which is used in the natural texts they are supposed to emulate. The deficiency is considerable. When the scientists use, on average, four different words, the best generator (SCIgen) uses three. In other words, the current generators do not seem able to mobilize as much vocabulary as the specialists in the field. Of course the

---

[1] SBIR (Small Business Innovation Research) is a program run by the US government.

Table 2 Vocabulary Richness

	Richness (for 10k tokens)	Standard deviation (tokens)	Corpus length (number of tokens)
Generated texts			
SCIgen	1,539	15.1	178,956
Mathgen	1,254	18.7	30,212
Scigen-physics	1,433	14.9	33,473
Propgen	603	3.1	26,603
Scientific NT			
Computer science	2,178	28.6	101,839
Political speeches			
Obama' State of the Union	2,022	25.5	40,771
Tony Blair speeches	2,277	33.2	1,524,071

software using Markov process is not affected by this limit because their lexicon is contained in the natural texts that comprise the training corpus.

3.2 Length and Structure of Sentences

The length and structure of sentences are indices of the stylistic choices of the author(s) of a text [32]. Table 3 summaries these stylistic characteristics of the corpora.

The adaptation of the SCIgen software to mathematics and physics was accompanied by shortening of the sentences. Within this limit, regardless of the field, not only are the chimera sentences too short, but, crucially the distribution of the sentence lengths in the GT is very different from that observed in NT. In the three corpora of

Table 3 Key figures of the sentence lengths in scientific GT compared to the natural ones (in word tokens)

	Mean length	Standard deviation	Modal length	Median length	Medial length
GT					
SCIgen	13.7	8.9	12	13.3	16.7
Mathgen	9.0	6.6	10	9.2	11.6
Scigen-physics	11.6	10.1	10	11.0	16.6
NT					
Computer science	17.3	13.4	1	16.4	23.1

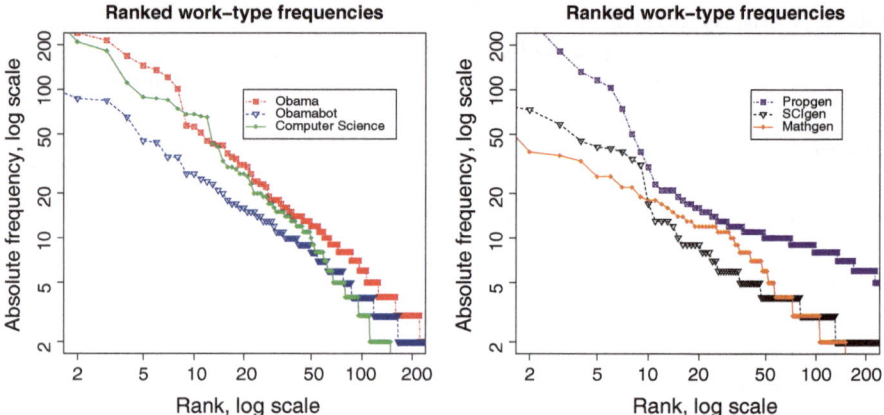

Fig. 2 Ranked-type frequency

chimerae (first part of the Table), the three central values (mean, median, and mode) are very close, that indicates a nearly Gaussian distribution (bell curve shape). In NT, this distribution is asymmetric (Mode < Median < Mean < Medial) and indicates the predominance of short sentences but also a wide range of lengths and the presence of rather long sentences. For example, in the *computer science* corpus, half of the texts are covered by sentences, the lengths of which are more than 23 word tokens (medial length).

3.3 Distribution of Word-Type Frequencies

Third, when ranking the word types of GT by ascending frequencies, the distribution is not what would be expected in a *n*atural text. According to the so-called "Zipf-law," if the texts are long enough, the natural distribution will follow more or less straight lines along the diagonal of the log–log diagram (left part of Fig. 2). The right part of Fig. 2 shows that the texts by the three generators are very far from this distribution with some jumps and thresholds at certain frequencies.

Because GT are not only meaningless but also formally very far from the NT they are supposed to emulate, they should be easy to detect. Yet, these texts have deceived many robots and, even, some people. In 2012–2013, more than a hundred fake papers by SCIgen were found in the IEEE Xplore bibliographic database and sixteen in the Springer'one [23, 35]. These papers were supposed to have been selected through a peer-review process under the supervision of scientific committees including senior academics.

4 Distance and Hierarchical Clustering

The fake papers in these bibliographic databases were detected with the help of an automatic procedure combining the calculation of the intertextual distance with automatic clustering.[2]

Intertextual Distance

The distance between two texts A and B is measured using the following method detailed in Appendix (see [23, 26]).

The distance varies evenly between 0—the same vocabulary is used in both texts (with the same frequencies)—and 1 (the texts do not share any word tokens). This distance between two texts can be interpreted as the proportion of different word tokens in both texts. A distance of $\delta_{(A,B)} = 0.4$ means that the two texts share 60% of their word tokens (without reference to token order in both texts). An intertextual distance of δ can be interpreted as follows: choosing randomly 100 word tokens in each text, δ is the expected proportion of common word tokens between these two sets of 100 words.

Intertextual distance depends on four factors. In order of decreasing importance, they are as follows: genre, author, subject, and epoch. An unusually small intertextual distance, between two texts in the same genre (e.g., computer science papers), suggests striking similarities and/or texts by the same author on the same topic.

Agglomerative Hierarchical Clustering

Properties of intertextual distance make it possible to establish agglomerative hierarchical clustering and graphical representations of the relative proximities between texts [30, 40].

This representations are used to identify more or less homogeneous groups within a large population. The clustering algorithm proceeds by grouping the two texts separated by the smallest distance and by recomputing the average (arithmetic mean) distance between all other texts and this new set, and so on until the establishment of a single set. These successive groupings are represented by a dendrogram with a scale representing the relative distances corresponding to the different levels of aggregation. By cutting the graph, as close as possible to a threshold considered as significant, one can distinguish groups of texts as very close, fairly close, etc. The higher the cut is made, the more heterogeneous the classes are and the more complex the interpretation of the differences is.

To correctly analyze these figures, it must be also remembered that whatever their position on the nonscaled axis, the proximity between two texts or groups of texts is measured by the height at which they are united.

All figures presented in the following are computed using the software R [12] and corpora presented in Table 4.

---

[2] Available online: http://scigendetection.imag.fr.

Table 4 Corpora

Corpus Name	Generator	Number of texts
SCIgen	SCIgen (PCFG)	12
Mathgen	Mathgen (PCFG)	11
Scigen-physics	Scigen-physics (PCFG)	12
Obamabot	Obamabot (Markov Chain)	2
Obama	Obama (Human)	5
CS	Computer science (Human)	12

GT Separated from NT

Figure 3 shows that the method actually identifies the texts generated by the different PCFG and separates them from the natural ones. As a matter of fact, classification clearly separates the GT and the NT.

For example, between the SCIgen texts and the natural ones, the average distance is 0.62, clearly outside the intervals of fluctuation for contemporaneous NT written in the same genre by different authors. Thus three obvious conclusions can be drawn. First, the SCIgen texts are very far from the NT they are supposed to imitate. Second, SCIgen is behaving as a single author who always faces the same subject in the same situation of utterance. Third, intertextual distance combined with automatic clustering offers an effective tool for specific authorship attribution: the detection of GT (as the generators behave like a single author).

The position of President Obama's State of Union addresses is very interesting. First, these texts are more or less grouped at the same level as the GT's edges that indicate not only the same authorship for all the addresses, but also the fact that the complex elaboration of these speeches is not so far from those that could have been produced by a generator! Second, the two *Obamabots* generated by software using Markov process are clustered with the texts they imitate. This result is not surprising since the real speeches constituted the training corpus. This highlights the obvious limitation of current procedures for GT detection (our own included): they work on the vocabulary (and frequencies of words) without dealing with the meaning of these texts (our conclusion discusses this problem). The main conclusion remains: the generators act as single authors dealing with the same themes and do not present the diversity of NT produced by different authors.

Most of the generated texts can be clearly distinguished from those written by humans, which is not the case with the Obamabot-generated texts.

5   ROUGE Measures

One of the most widespread evaluation approaches in NLP uses automatically computed metrics measuring *n*-gram coverage between a candidate output and one or several reference texts. In particular, in the domain of automatic summarisation,

Fig. 3 Clustering each type of texts using intertextual distance

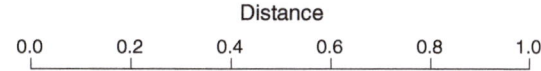

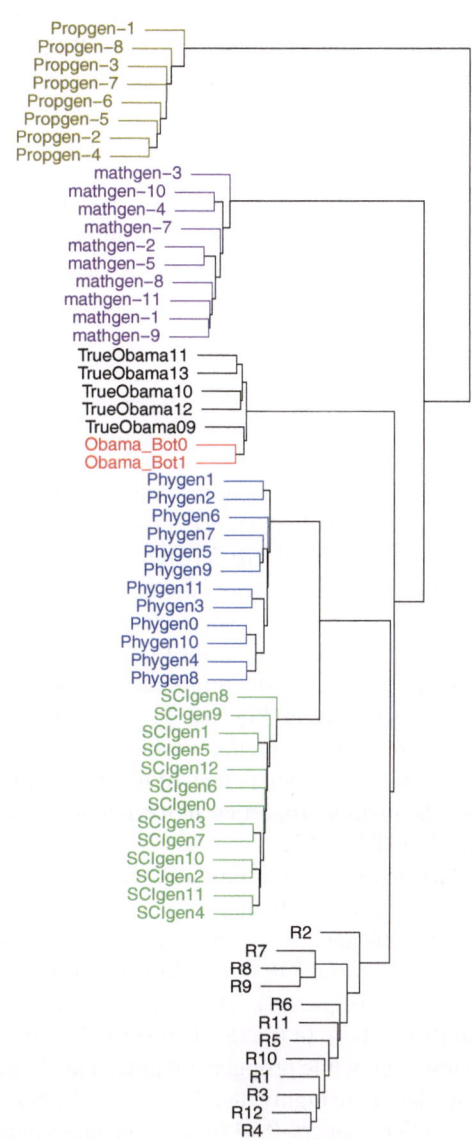

ROUGE (Recall-Oriented Understudy for Gisting Evaluation) [29] has become a reference method to evaluate summarisation systems (see the DUC and TAC conferences). Moreover, ROUGE is also a method used in the evaluation of some NLG system outputs [13] although it is unclear how well its results correlate with human evaluations [38]. ROUGE was developed to compare automatic and human-authored extractive summaries. Here extractive means that the generated summary is composed only from the material "extracted" from the original source(s). In short the summaries are—sometimes slightly modified—"extracts" of the main texts. Roughly speaking, the metric aims at assessing how well the candidate summary covers the reference summaries using the frequency of the words and the sequence of the words. Thus ROUGE reflects the similarity between a GT and the gold standard reference. Therefore, we expect GT (resp. NT) to have a high ROUGE score within each group (i.e., high similarity).

ROUGE is not a unique measure but a set of measures whose simplest family is ROUGE-N which measures n-gram overlap. Equation (3) shows how ROUGE-N $\in [0, 1]$ is computed. Basically, it is the ratio of n-gram in the reference that are also found in the candidate. This is thus a recall measure.

$$\text{ROUGE-N}(ref, cand) = \frac{\sum_{S\in\{ref\}} \sum_{gram_n\in S} Count_{match}(gram_n, cand)}{\sum_{S\in\{ref\}} \sum_{gram_n\in S} Count(gram_n)} \quad (3)$$

where $Count_{match}(gram_n, cand)$ is the number of occurrences of a n-gram of the reference in the candidate. In the present paper, we set n to 3 so ROUGE-1 (unigram), ROUGE-2 (bigram), and ROUGE-3 (trigram) were computed. We believe that these will be particularly adapted to detect the texts generated using Markov chain models and the texts that share the same vocabulary (unigram, bigram).

ROUGE-L was also used. It finds the LCS(X,Y), the longest common subsequence between the candidate X and the reference Y. The recall of this measure R_{LCS} for a candidate is computed using $R_{LCS} = \sum_{s\in R} LCS \cup (s, C)/m$, where R is the set of sentences of the reference which contains m words and C the set of sentences of the candidate. The hypothesis is that ROUGE-L will be higher in GT by a PCFG since some of the textual properties at document and sentence levels would be more recurrent than in GT by MC.

The ROUGE measures were computed using the ROUGE package [28] on the texts of the corpora (cf. Sect. 2.2). Each pair of text was successively used as candidate and reference and the distance was computed using $1 - \text{F-measure}(ROUGE)$. Figure 4 shows the results for ROUGE-1 and ROUGE-L on the raw texts (Fig. 4.a, b) and without stop words (Fig. 4c, d). The measure enables a successful identification of clusters of uniform class. In the raw text case, the three scientific Corpora by PCFG are grouped together, while propgen, Obama, and NT scientific texts constitute other groups. But the latter are highly discriminated. The NT scientific papers are at around 0.75 from the SCIgen papers, 0.74 from the propgen papers, and 0.65 from the Obama speeches. Hence ROUGE seems to be a very efficient way of discriminating SCIgen from non-SCIgen papers. ROUGE-2 and ROUGE-3 measures were also computed but although they grouped the texts perfectly the measure intergroup was much higher which made the discrimination much more difficult. This would suggest that

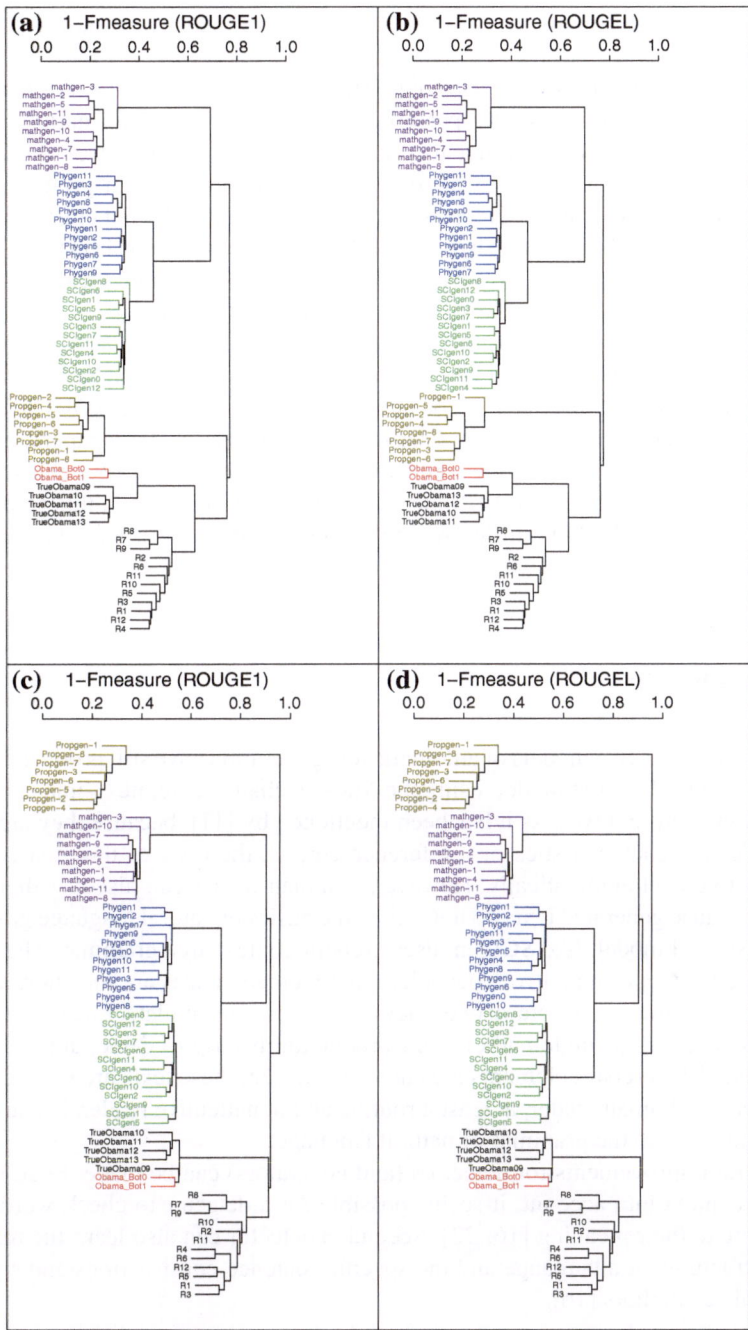

Fig. 4 Clustering each type of texts using $1 - \text{F-measure}(ROUGE)$ as distance with and without stop words. **a** ROUGE-1 raw texts. **b** ROUGE-L raw texts. **c** ROUGE-1 without stop words. **d** ROUGE-L without stop words

GT contains a high amount of variation in sequence of tokens but not in vocabulary (cf. Sect. 3.3).

The ROUGE measures were also applied to texts from which stop words have been removed. This processing is often performed for some types of Information retrieval and summarisation tasks in order to process only words that convey meaningful information.[3] The effect was to increase all the measures within groups making the discrimination between SCIgen and non-SCIgen a bit easier but clustering more scatered. This would support the hypothesis that stop words play a role in the signature of the SCIgen papers [15]. However, the propgen texts are more clearly excluded from the other texts which make this preprocessing quite interesting.

Overall, these results confirm the findings of the previous distance. (1) The NT are close to each other (Obama and scientific NT) while (2) the SCIgen papers are grouped together and propgen is considered as satellite. (3) The ROUGE metrics seem to be a very effective way of discriminating between SCIgen papers, propgen texts, and NT. However, since they are based on frequency they are not able to detect Obama_Bot papers. Even the LCS measure was not informative in this respect but this might be a result of overfitting when the language model was learned. This calls for further investigations.

6   Conclusion

In conclusion, the two models of automatic text generation have still produced inconclusive results. The first model, using the Markov chains, emulates some characteristics of the natural texts. As it has been mentioned by [11], because they are based on the lexical characteristics of a reference corpus, the texts of this first type are difficult to detect automatically. However, a human reader can discover them very easily, because generated texts do not follow the basics of natural language grammar.

The second model, like SCIgen, uses prebuilt context-free grammar. The generated texts have poor repetitive vocabulary and their sentences are too short and too uniform compared to the natural ones they are supposed to emulate. Each one of these generators acts as a single author, their production is easy to detect automatically. However, if the prebuilt elements are carefully chosen, these texts are more difficult to detect by a human reader, at least a routine and nonattentive reader, because they are conforming to the grammar of natural language.

Several improvements for detectors (and generators) can be made. First, by taking the context into account, it seems possible for a detector to check word usage according to their meaning [16, 22]. Second, a detector can also learn the real syntactic structures of a language and the specific sentence constructions and styles of the emulated authors [16].

---

[3]This preprocessing is not systematic. For instance, the lasts DUC conferences did include the stop words in the ROUGE computing.

These area of research will be of some help not only for fake paper detection, but also against plagiarism, duplication, and other malpractices. More importantly, these improvements could be of great help regarding both understanding and generating natural texts. They will help improve software for processing and generating texts, for data scientists, lexicographers, translators, language teachers, and all users of large digital text and data bases.

Acknowledgments The authors would like to thank Edouard Arnold (Trinity College Dublin) for his valuable reading of previous versions of this paper as well as the organizers of the Flow Machines Workshop 2014 among which are François Pachet, Mirko Degli Esposti, Vittorio Loreto, and Eduardo G. Altmann.

Appendix

Given two texts A and B, let us consider

- N_A and N_B: the number of *word tokens* in A and B, respectively, i.e., the lengths of these texts;
- F_{iA} and F_{iB}: the absolute frequencies of a type i in texts A and B, respectively;
- $|F_{iA} - F_{iB}|$ the absolute difference between the frequencies of a type i in A and B, respectively;
- $D_{(A,B)}$: the intertextual distance between A and B is as follows:

$$D_{(A,B)} = \sum_{i \in (A \cup B)} |F_{iA} - F_{iB}| \quad with \quad N_A = N_B \tag{4}$$

The distance index (or relative distance) is as follows:

$$D_{rel(A,B)} = \frac{\sum_{i \in (A \cup B)} |F_{iA} - F_{iB}|}{N_A + N_B} \tag{5}$$

If the two texts are not of the same lengths in tokens ($N_A < N_B$), B is "reduced" to the length of A:

- $U = \frac{N_A}{N_B}$ is the proportion used to reduce B in B'
- $E_{iA(u)} = F_{iB}.U$ is the theoretical frequency of a type i in B'

In Eq. (4), the absolute frequency of each word type in B is replaced by its theoretical frequency in B':

$$D_{(A,B')} = \sum_{i \in (A \cup B)} |F_{iA} - E_{iA(u)}|$$

Putting aside rounding-offs, the sum of these theoretical frequencies is equal to the length of A. Equation (5) becomes

$$D_{rel(A,B)} = \frac{\sum_{i \in (A \cup B)} |F_{iA} - E_{iA(u)}|}{N_A + N_{B'}}$$

References

1. Arnold, E.: Le discours de tony blair (1997–2004). Corpus **4**, 55–77 (2005)
2. Arnold, E.: Le sens des mots chez tony blair (people et europe). In: H. Serge, P. Bénédicte (eds.) 9e Journées internationales d'analyse statistique des données textuelles, vol. 1, pp. 109–119. Presses universitaires de Lyon (2008)
3. Ball, P.: Computer conference welcomes gobbledegook paper. Nature **434**, 946 (2005)
4. Balpe, J.P.: Fiction et écriture générative. les Actes de Lecture (103), 37–48 (2008)
5. Barbieri, G., Pachet, F., Roy, P., Esposti, M.D.: Markov constraints for generating lyrics with style. ECAI **242**, 115–120 (2012)
6. Baughn, J.: http://nonsense.sourceforge.net (2001). Online; Accessed 11 December 2014
7. Chomsky, N.: Three models for the description of language. IEEE Trans. Inf. Theory **2**(2), 113–124 (1956)
8. Dalkilic, M.M., Clark, W.T., Costello, J.C., Radivojac, P.: Using compression to identify classes of inauthentic texts. In: Proceedings of the 2006 SIAM Conference on Data Mining (2006)
9. Doug, C., Jan, P., Penelope, S.: A practical part-of-speech tagger. In: ANLC '92 Proceedings of the Third Conference on Applied Natural Language, pp. 133–140 (1992)
10. Elmacioglu, E., Lee, D.: Oracle, where shall i submit my papers? Commun. ACM (CACM) **52**(2), 115–118 (2009)
11. Fahrenberg, U., Biondi, F., Corre, K., Jégourel, C., Kongshøj, S., Legay, A.: Measuring Structural Distances Between Texts. CoRR arXiv:1403.4024 (2014)
12. Feinerer, I., Hornik, K., Meyer, D.: Text mining infrastructure in r. J. Stat. Softw. **25**(5), 1–54 (2008)
13. Gatt, A., Portet, F.: Textual properties and task-based evaluation: Investigating the role of surface properties, structure and content. In: 6th International Conference on Natural Language Generation (INLG-10) (2010)
14. Georg, B.: https://bitbucket.org/birkenfeld/scigen-physics (2014). Online; Accessed 11 December 2014
15. Ginsparg, P.: Automated screening: arxiv screens spot fake papers. Nature **508**(7494), 44–44 (2014). http://dx.doi.org/10.1038/508044a
16. Halliday, M.A.K., Webster, J.: Computational and quantitative studies. In: Halliday, M.A.K., Webster, J. (eds.) Continuum. London, New York (2004)
17. Hirsch, J.E.: An index to quantify an individual's scientific research output. Proc. Natl. Acad. Sci. **102**, 16569–16572 (2005)
18. Hockey, S., Martin, J.: OCP Users' Manual. Oxford University Computing Service, Oxford (1988)
19. Hubert, P., Labbé, D.: Vocabulary richness. In: Communication au congrès de l'ALLC-ACH. Paris: La Sorbonne. Reproduced in Lexicometrica, 1997 (1994)
20. John, S.: Corpus, Concordance, Collocation. Oxford University Press, Oxford (1991)
21. Labbé, C.: Ike antkare, one of the great stars in the scientific firmament. Int. Soc. Scientometr. Informetr. Newsl. **6**(2), 48–52 (2010)
22. Labbé, C., Labbé, D.: How to measure the meanings of words? amour in corneille's work. Lang. Resour. Eval. **39**(4), 335–351 (2005)
23. Labbé, C., Labbé, D.: Duplicate and fake publications in the scientific literature: how many scigen papers in computer science? Scientometrics **94**(1), 379–396 (2013)

24. Labbé, C., Labbé, D.: Was shakespeare's vocabulary the richest? In: Proceedings of the 12th International Conference on Textual Data Statistical Analysis, pp. 323–336. Paris (2014)
25. Labbé, C., Portet, F.: Towards an abstractive opinion summarisation of multiple reviews in the tourism domain. In: SDAD 2012, The 1st International Workshop on Sentiment Discovery from Affective Data, pp. 87–94 (2012)
26. Labbé, D.: Experiments on authorship attribution by intertextual distance in english. J. Quant. Linguist. **14**(1), 33–80 (2007)
27. Lavoie, A., Krishnamoorthy, M.: Algorithmic Detection of Computer Generated Text. ArXiv e-prints (2010). arXiv:1008.0706
28. Lin, C.Y.: Rouge: A package for automatic evaluation of summaries. In: Text Summarization Branches Out: Proceedings of the ACL-04 Workshop, pp. 74–81 (2004)
29. Lin, C.Y., Hovy, E.: Automatic evaluation of summaries using n-gram co-occurrence statistics. In: Proceedings of HLT-NAACL-03, pp. 71–78 (2003)
30. Manning, C.D., Raghavan, P., Schütze, H.: Introduction to Information Retrieval. Cambridge University Press, New York (2008)
31. Manurung, R., Ritchie, G., Pain, H., Waller, A., O'Mara, D., Black, R.: The construction of a pun generator for language skills development. Appl. Artif. Intell. **22**(9), 841–869 (2008)
32. Monière, D., Labbé, C., Labbé, D.: Les styles discursifs des premiers ministres québécois de jean lesage à jean charest. Revue canadienne de science politique **41**(1), 43–69 (2008)
33. Nadovich, C.: Automatic SBIR proposal generator, http://www.nadovich.com/chris/randprop/ (2014). Online; Accessed 11 December 2014
34. Nathaniel, E.: http://thatsmathematics.com/mathgen/ (2012). Online; Accessed 11-December-2014
35. Noorden, R.V.: Publishers Withdraw more than 120 Gibberish Papers. Nature (24 February 2014)
36. Parnas, D.L.: Stop the numbers game. Commun. ACM **50**(11), 19–21 (2007)
37. Portet, F., Reiter, E., Gatt, A., Hunter, J., Sripada, S., Freer, Y., Sykes, C.: Automatic generation of textual summaries from neonatal intensive care data. Artif. Intell. **173**(7–8), 789–816 (2009)
38. Reiter, E., Belz, A.: An investigation into the validity of some metrics for automatically evaluating natural language generation systems. Comput. Linguist. **35**(4), 529–558 (2009)
39. Reiter, E., Dale, R.: Building Natural Language Generation Systems. Studies in Natural Language Processing. Cambridge University Press, Cambridge (2000)
40. Sneath, P., Sokal, R.: Numerical Taxonomy. Freeman, San Francisco (1973)
41. Stribling, J., Krohn, M., Aguayo, D.: Scigen, http://pdos.csail.mit.edu/scigen/ (2005). Online; Accessed 11 December 2014
42. Stubbs, M.: Texts and Corpus Analysis. Blackwell, Oxford (1996)

Universality of Stylistic Traits in Texts

Efstathios Stamatatos

Abstract The style of documents is an important property that can be used as discriminant factor in text mining applications. Among the great number of possible measures proposed to quantify writing style there are some features that can be characterized as universal, in the sense that they can be easily extracted from any kind of text in practically any natural language and provide accurate results when used in style-based text categorization tasks. In this paper we examine whether such universal stylometric features remain effective under difficult scenarios where the topic and/or genre of documents used in the training phase differ from that of the questioned documents. Based on a series of experiments in authorship attribution, we demonstrate that character n-gram features are reliable and effective given that the appropriate number of features is used. It is also shown that when the number of candidate authors increases, the representation dimensionality should also increase to improve classification results.

1 Introduction

Large amounts of electronic texts are produced daily, a great part of which is available online through Internet services. As a consequence, the need to handle textual information efficiently is now greater than ever. A large body of research in text mining attempts to develop methodologies and build tools performing text categorization and filtering, text clustering, text summarization, etc. [40]. Documents can be described by several factors and properties. The most prevalent factor is their topic or theme. This can be used to build document taxonomies or filter texts according to their topic [32]. Another important property is the sentiment of texts, especially important when one attempts to handle the vast amount of opinionated texts available in social media [28].

E. Stamatatos (✉)
University of the Aegean, Karlovassi, Greece
e-mail: stamatatos@aegean.gr

© Springer International Publishing Switzerland 2016   143
M. Degli Esposti et al. (eds.), *Creativity and Universality in Language*,
Lecture Notes in Morphogenesis, DOI 10.1007/978-3-319-24403-7_9

Another factor that characterizes texts is their style. Although hard to define exactly what style is, there are two main aspects of style especially useful for distinguishing between texts:

- Functional style: this depends on the functional purpose of the texts, it is strongly associated with its form and the medium used to publish the texts as well as their genre and register. For example, we expect that all research papers share some stylistic choices no matter what the topic of the research is or who the author is. The same is true for email messages, newspaper articles, blogs, etc. The task of attempting to exploit this type of writing style is called *genre identification* and has important applications in information retrieval and natural language processing [15, 21, 26, 30].
- Authorial style: this depends on the author of the texts and is composed of their personal use of language and idiosyncrasies. We expect that all texts written by the same author share some stylistic choices not affected by the topic and the genre of texts. A great number of research studies have been performed in *authorship attribution* (also called authorship identification) attempting to reveal the authors of anonymous documents or to solve disputed cases where several individuals claim the authorship of a certain document [10, 19, 20, 23, 37]. Significant forensic applications are associated with this task [1]. We also expect that groups of authors that share some demographics (i.e., age, gender, education, etc.) would have striking similarities in their personal writing style. This led to the recent development of the *author profiling* community that attempts to extract characteristics of authors from analyzing their texts [29].

For any given document, the combination of its topic, functional and authorial styles produce a unique blend that may be considered as a *document fingerprint* and can be used to identify cases of plagiarism or text reuse (i.e., when parts of one document have been used in another document) [36]. In such cases, stylistic inconsistencies may be especially useful to detect the suspicious parts of documents [35].

In order to be able to use writing style in the framework of a text mining application, we need to quantify it. The line of research dealing with the quantification of style is called *stylometry* and has a long history. The first pioneering studies in stylomtery, dating back to the nineteenth century, were based on the quite laborious task of manual counting word or letter frequencies in long documents [25, 41]. Later on, the availability of computational systems and tools enabled researchers to use automated analysis and extract rich sets of features that can describe the style of texts. Examples of such features are vocabulary richness measures, frequencies of function words, character n-grams, parts of speech, etc. [15, 17, 23, 39].

In general, to achieve the best possible results, researchers use a combination of stylometric features that is suitable for a particular case [11, 39]. Many features are application-specific and can only be used when the examined texts share some properties. For example, if it is known that all documents belong to a certain genre, then appropriate genre-specific measures can be defined (e.g., the use of greetings in email messages) [9], if it is known that all documents are in the same thematic area, then appropriate topic-specific measures can be used [42]. In addition, the language of documents may be an obstacle toward the application of sophisticated natural language processing tools which are able to extract syntactic or semantic-related stylometric features [3, 10, 37]. On the other hand, there are certain types of features that may be considered *universal* since they can be used in any case (i.e., practically all kinds of genres and natural languages).

In this paper, we examine two types of such universal stylometric features: function words and character n-grams. They can easily be extracted from every document and they have been successfully used in several style-based text categorization tasks like authorship attribution, automatic genre identification, and plagiarism detection. Focusing on the authorship attribution task, we aim at examining the effectiveness of these features in difficult cases where there are differences in topic and genre of the documents under examination. It is demonstrated that one crucial decision concerns the representation dimensionality and it is possible to attain high accuracy results given that the appropriate number of features is used. Moreover, we show that the number of candidate authors also affects this decision, especially for character n-gram features.

The rest of the paper is organized as follows. Section 2 presents several approaches to the quantification of style. Section 3 discusses authorship attribution tasks and defines challenging scenarios for universal features while Sect. 4 presents experimental settings and results. Finally, Sect. 5 summarizes the main conclusions of this study.

2 Stylometry

There is not a consensus on the definition of style. As a consequence, a lot of different approaches have been reported in previous studies aiming at quantifying some textual properties considered to be associated with stylistic choices. An excellent review of early stage stylometry is presented by [12] while a more recent survey is given by [34]. In this section, we describe the basic categories of stylometric features found in style-based text categorization tasks, mainly authorship attribution and genre identification.

The most commonly used type of information used in stylometric studies refers to lexical features. Text can be seen as a sequence of words grouped in sentences. Thus, word length, sentence length, number of unique words (hapax legomena), type/token ratio, and other vocabulary richness measures were very popular in early stage studies [25, 41]. Another popular approach is to use word frequencies, especially of very frequent closed-class words, like articles, prepositions, conjunctions, etc., also known as *function words*. Such words are very important since they are usually associated with certain syntactic structures so their frequency is an indirect measure of syntactic information. The set of function words may be predefined for all tasks [1, 2], or specifically chosen for a given task [27], or extracted automatically for any given task using the most frequent words of a training corpus [6].

Another popular idea is to use character features. According to this approach, texts can be seen as strings of characters. Frequencies of letters, digits, punctuation marks, etc., belong to this category [9, 42]. Moreover, character sequences, like prefixes and suffices provide an indirect measure of lexical and syntactic information [24]. A simple approach that has been proved very effective in many tasks is based on the set of most frequent sequences of characters, known as character n-grams [11, 15, 17, 23]. This method is able to capture many types of information (lexical, syntactic, formatting, etc.) and does not require complicated tools for extracting the relevant measures. Compression-based methods (using text compression algorithms as a means to measure stylistic homogeneity) also exploit information from character sequences [5, 16].

In theory, syntactic structures and semantic forms provide more reliable stylistic information since they should not be affected by topical shifts and are used by the authors unconsciously [3, 10, 22, 33, 39]. However, their use requires the availability of certain natural language processing tools than can analyze the documents within a task and provide accurate syntactic or semantic measures. Therefore, such features are language dependent (they can be used only for languages where appropriate tools are available) and noisy (the tools make errors and the provided measures may not be 100 % correct). The most popular features of this category are parts-of-speech frequencies mainly because parts-of-speech tagging is effectively performed by existing tools in many languages.

Another source of stylistic information is the layout or the presentation of the document. This is especially useful in genre identification since certain genres are strongly associated with specific document layouts (e.g., research papers may be multicolumn with tables, graphs, etc.). In the case of web pages, such structural features can easily be extracted (e.g., HTML tag and meta tag frequencies, image counts, use of JavaScript, number of links, etc.) and their use together with textual features increases the potential of the stylometric model [15, 21, 26, 30]. However, they are not general-purpose features since the format of documents within a certain task may not provide such information.

To take advantage of certain properties of the available documents within a task, other application-specific features may be defined. Mainly, they attempt to exploit the fact that all documents are matched for genre (e.g., the use of greetings and farewells in email messages) [9], or topic (e.g., the use of certain topic-specific words, like deal or sale, in texts about computer sales) [42], or language (e.g., use of slang words in conversations) [8]. This type of information is especially useful since it permits the stylometric model to be adapted to the properties of a specific set of documents.

In general, the combination of several types of features increases the effectiveness of the resulting model [11, 15, 39]. Another idea is to attempt to use the most appropriate type of features according to the properties of a certain case [31]. Moreover, in the vast majority of published studies, for each feature a single measure is extracted from a text. Alternatively, distributional measures indicate how a certain feature varies within a text [13]. From another perspective, graph-based models have been proposed to capture dependencies between different features [4].

3 Universality in Authorship Attribution

Authorship attribution may be viewed as a single-label multiclass text categorization problem. In general, we are given a set of candidate authors and for each one of them we get undisputed samples of their texts. This is the training corpus that can be used to build a model able to distinguish between the text samples of candidate authors. Then, any document of disputed authorship may be assigned to one of the candidate authors using this model. There are three main tasks in authorship attribution:

- Closed-set attribution: where the true author of any disputed text is necessarily one of the candidate authors. This is the easiest case and most of the studies in authorship attribution follow this scenario [11, 17, 23, 37]. It is appropriate for most forensic applications where police investigators are able to limit the number of suspects based on evidence about their knowledge of certain issues or their accessibility to certain resources.
- Open-set attribution: where the true author of a disputed text may not be included in the set of candidate authors. This is the most general scenario and resembles any case where it is not possible to limit the number of suspects. Previous work have shown that this task is more difficult when the set of candidate authors is small (2 or 3) rather than large [20].
- Author verification: where the set of candidate authors is singleton [19]. Any authorship attribution problem, either closed-set or open-set, can be decomposed into a set of author verification cases. Therefore, the ability to solve this problem is of crucial importance and there is increasing interest on this task recently [18, 38].

The vast majority of published studies in authorship attribution only consider the case where all texts in both training corpus (documents of known authorship) and evaluation corpus (documents of unknown or disputed authorship) are in the same thematic area and belong to the same genre. For many practical applications these assumptions seem reasonable. However, there are certain cases where such assumptions do not hold. For example, we can imagine one scholar attempting to verify the authenticity of a suicide note requiring the availability of other samples of suicide notes from all suspects [7]. It is therefore crucial, at least for certain applications, the stylometric method we use to remain effective even when the available documents are not matched for topic or genre. Thus, we define four scenarios for examining the robustness of an authorship attribution model:

- Same topic same genre: the simplest case, where the training texts we use to build the model and the disputed texts are in the same thematic area and belong to the same genre.
- Cross-topic same genre: where there are differences in the topic of training and disputed texts while they all belong to the same genre.
- Same topic cross-genre: where the training and disputed texts are in the same thematic area but differ in genre.

- Cross-topic cross-genre: where the training and disputed texts do not agree in topic and genre, certainly the most difficult case.

As already discussed, stylometric approaches based on function words and character n-grams may be considered as universal given that they can be easily applied to any kind of texts and practically all natural languages. Moreover, they have produced competitive performance results in previously published studies [11, 17, 19, 20, 23]. However, it remains to be seen how much they are affected under cross-topic or cross-genre conditions. The remainder of this paper deals with this question.

4 Experiments

4.1 Corpus

The corpus used in this study is composed of texts published in *The Guardian* daily UK newspaper. The texts were downloaded using the publicly available API[1] and preprocessed so that to keep the unformatted main text (titles, name of authors, dates, tags, images, etc., were removed).

The opinion articles of this newspaper (comments) are described using a set of tags indicating their subject. There are eight top-level tags (World, US, UK, Belief, Culture, Life&Style, Politics, Society) each one of them having multiple sub-tags. It is possible (and very frequent) for an article to be described by tags belonging to different main categories (e.g., a specific article may belong to all UK, Politics, and Society). In order to have a clearer picture of the thematic area of the collected texts, in the presented corpus we only used articles that belong to a single main category. Therefore, each article can be described by multiple tags all of them having to belong to a single main category. Moreover, articles coauthored by multiple authors were discarded.

In addition to opinion articles on several thematic areas, this corpus comprises book reviews, a different genre. Book reviews are also described by a set of tags similar to the opinion articles. However, no thematic tag restriction was taken into account when collecting book reviews.

Table 1 shows details about this corpus. It comprises texts from 13 authors selected so that they have published texts in multiple thematic areas (*Politics, Society, World, UK*) and different genres (opinion articles and book reviews). At most 100 texts per author and category have been collected, all of them published within a decade (from 1999 to 2009). Note that the opinion article thematic areas can be divided into two pairs of low similarity, namely *Politics-Society* and *World-UK*. In other words, the *Politics* texts are more likely to have some thematic similarities with *World* or *UK* texts rather than with *Society* texts.

---

[1] http://explorer.content.guardianapis.com/.

Table 1 The corpus used in this study comprising documents in different topics and genres

Author	Politics	Society	World	UK	Books reviews
CB	12	4	11	14	16
GM	6	3	41	3	0
HY	8	6	35	5	3
JF	9	1	100	16	2
MK	7	0	36	3	2
MR	8	12	23	24	4
NC	30	2	9	7	5
PP	14	1	66	10	72
PT	17	36	12	5	4
RH	22	4	3	15	39
SH	100	5	5	6	2
WH	17	6	22	5	7
ZW	4	14	14	6	4
Total	254	94	377	119	160

4.2 Experimental Settings

Unfortunately, it is not possible to examine all four scenarios mentioned at the end of Sect. 3 using the Guardian corpus. In particular, it is not possible to examine the last two (intra-topic cross-genre and cross-topic cross-genre) since there is limited information about the topic of the available book reviews. Therefore, we merge these two cross-genre scenarios into one. We focus on closed-set attribution as described in Sect. 3. In each author identification experiment all training texts are opinion articles of a certain topic while all evaluation texts come either from the same topic (same topic, same genre), a different topic (cross-topic, same genre), or a different genre (cross-genre scenario). Each time, at most 10 training/evaluation texts per author are used. When the training and evaluation sets come from the same category (same topic), training and evaluation texts are disjoint. Note that the training and evaluation texts are unevenly distributed among the candidate authors and this distribution varies according to topic or genre.

Two types of universal features are examined, namely words and character 3-grams. In both cases, the features are selected according to their total frequency of occurrence in the training corpus. Let V be the vocabulary of the training corpus (the set of different words or character n-grams) and $F = \{f_1, f_2, \ldots, f_v\}$ be the frequency of occurrence of all possible features in the training corpus. Given a predefined threshold t, we include in the feature set all features with $f_i \geq t$. The higher the t, the lower the dimensionality and vice versa. Therefore, it is possible to examine different sizes of the feature set by modifying t. In this study, the following frequency threshold values were used: 500, 300, 200, 100, 50, 30, 20, 10, 5, 3, 2, 1.

Note that for high values of frequency threshold and word features, we practically get function words only. As the frequency threshold gets lower more nouns, adjectives, and verbs are included in the list.

The well-known *support vector machines* classifier [14] is used in the experiments. This model can handle high-dimensional and sparse data, like the stylometric features we extract from texts, and it is considered one of the best algorithms for text categorization tasks. The linear kernel is used since the dimensionality of the representation is usually high (hundreds or thousands of features).

4.3   Results

Figures 1 and 2 show the micro-average classification accuracy results of attribution models based on word and character n-gram features, respectively, varying the representation dimensionality. It is obvious that character n-gram models are more effective than word-based models even in difficult cross-topic and cross-genre cases. In the simple case of same topic, same genre, character n-grams provides perfect classification results when the dimensionality is maximized. On the other hand, when topic or genre change, the appropriate selection of the dimensionality seems crucial for obtaining good performance. In those hard cases, performance drops after a certain point (about 3,000–4,000 features) when increasing the dimensionality. This drop is more dramatic for *Society* texts which may be considered as an opposite topic with respect to *Politics*. This indicates that in cross-topic cases, when topics significantly differ, lower dimensionality is advisable, since low-frequency features correspond to topic-specific information. The cross-genre case (*Book reviews*) seems not to be influenced so much by such topic-specific features. However, one should keep in mind that book reviews and opinion articles have many similarities and many of the book reviews included in the corpus talk about politics.

Word features produce the same general picture. The main difference is that performance drops much earlier, at about 500–1,500 features corresponding mainly to function words and some very frequent open-class words (nouns, verbs, etc.) The inclusion of more topic-specific words harms the word-based attribution models especially in cross-topic conditions where the topic significantly differs with that of the training texts (e.g., *Politics* vs. *Society*).

Next, we repeat the above experiment with a varying number of candidate authors. In more detail, we tested candidate set sizes of 2, 3, 5, and 8. For each candidate set size, 30 repetitions were performed by selecting (without replacement) a subset of the 13 authors included in the corpus. Figures 3 and 4 show the classification performance (averaged over the repetitions) for word and character n-gram features, respectively, in a same genre cross-topic scenario (training texts come from *Politics* topic while evaluation texts come from *World* topic) for varying candidate set sizes (the case where all 13 authors are included is also shown). Naturally, when the candidate set size grows larger classification performance drops (recall this is a closed-set classification task). Again, character n-gram models are more effective in

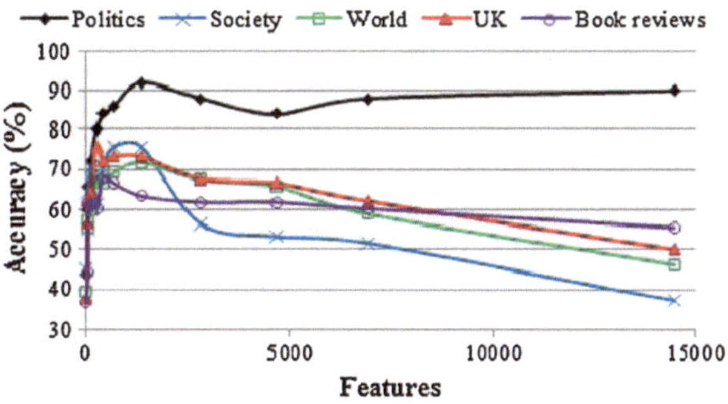

Fig. 1 Performance of the word-based attribution model trained on texts from Politics and evaluated with texts from the same topic (Politics), a different topic (Society, World, UK), or a different genre (Book reviews)

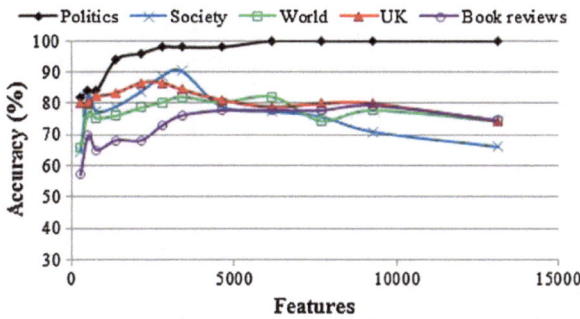

Fig. 2 Performance of the character 3-gram attribution model trained on texts from Politics and evaluated with texts from the same topic (Politics), a different topic (Society, World, UK), or a different genre (Book reviews)

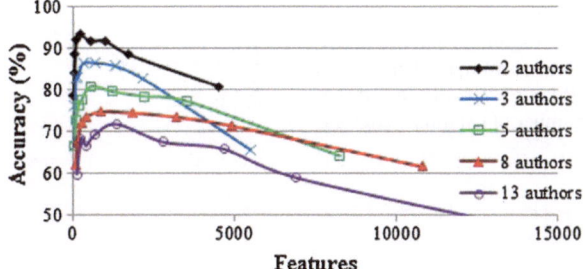

Fig. 3 Performance of the word-based attribution model trained on texts from one topic (Politics) and evaluated on texts from another topic (World) for a varying number of candidate authors

comparison to the respective word-based models. An interesting point is that, in both cases, the most appropriate dimensionality seems to depend on the candidate set size. The larger the candidate set size, the larger the number of features corresponding to the best obtained results. For character n-gram features, the optimal point seems to start at about 2,000 features for 2 candidate authors and grows to about 6,000 features for 13 authors. For word features, the optimal point starts at about 250 features for 2 authors and increases to about 1,500 features for 13 candidate authors.

Figures 5 and 6 show the results of a similar experiment, where the training texts come from opinion articles on *Politics* and evaluation texts come from a different genre (*Book reviews*). Again, we note that the number of character n-gram features corresponding to the best results starts at about 3,000 for 2 authors and gradually increases with the candidate set size reaching about 9,000 features for 13 authors.

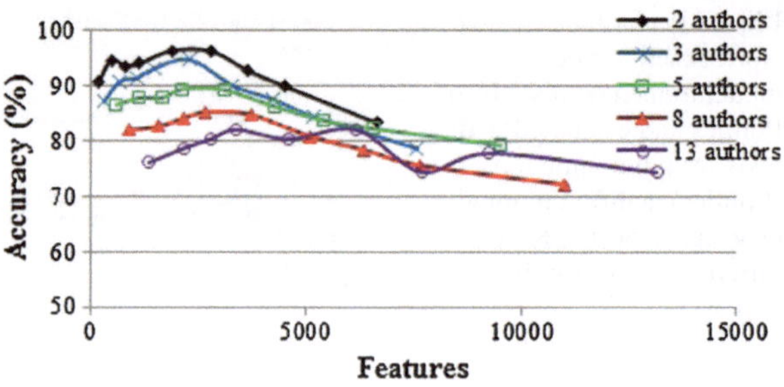

Fig. 4 Performance of the character 3-gram attribution model trained on texts from one topic (Politics) and evaluated on texts from another topic (World) for a varying number of candidate authors

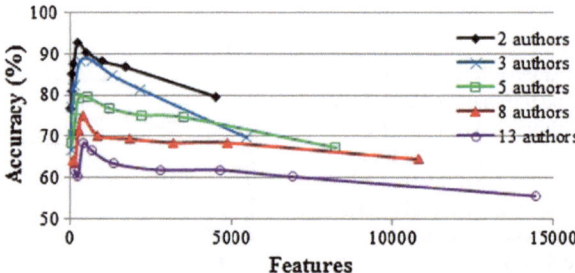

Fig. 5 Performance of the word-based attribution model trained on texts from one genre (opinion articles about Politics) and evaluated on texts from another genre (Book reviews) for a varying number of candidate authors

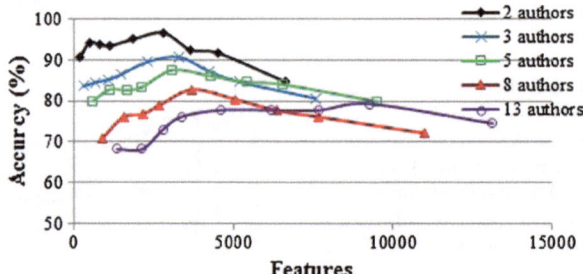

Fig. 6 Performance of the character 3-gram attribution model trained on texts from one genre (opinion articles about Politics) and evaluated on texts from another genre (Book reviews) for a varying number of candidate authors

In comparison to the previous cross-topic experiment, a higher number of features is required. This can be explained by the fact that many book reviews of this corpus talk about politics, therefore low-frequency topic-specific features are useful. On the other hand, it is noted that the number of word-based features that provide the best achieved performance does not vary that much. It starts at about 250 features for 2 candidate authors and reaches about 400 features for 13 authors. Since the most frequent words correspond to function words, these results indicate that function words are reliable features in cross-genre conditions.

On the other hand, word features seem unable to exploit the topic similarities (captured by low-frequency words) in cross-genre cases, in contrast to character n-gram features.

5 Conclusion

In this paper, we applied some well-known stylometric approaches, that is, function words and character n-grams to authorship attribution tasks. Such features are universal since they are easily available for practically any type of text and natural language. Beyond the case where all documents are matched for topic and genre, we examined their effectiveness under more challenging scenarios where the training and evaluation documents talk about different topics or belong to different genres. It is demonstrated that the attribution models, especially the ones based on character n-gram features, can be surprisingly effective in cross-topic or cross-genre conditions.

Character n-gram features performed better than word-based features in all experiments. The attribution models based on character n-grams require considerably higher dimensionality and are able to take advantage of low-frequency features where there are topic similarities among texts. On the other hand, word-based models mainly exploit topic-independent function words while low-frequency words seem to harm their effectiveness when there are changes in topic or genre.

One crucial decision concerns the dimensionality of the representation. It was shown that changes in topic and/or genre as well as the number of candidate authors considerably affect the appropriate choice of the number of features in the attribution models. In the simple scenario where training and evaluation documents are matched for topic and genre, maximum dimensionality is advisable for character n-gram features. When the topic or genre changes, the representation dimensionality should be carefully defined taking into account the number of candidate authors. However, it is not yet clear how this could be done formally.

It is not claimed that the examined models are the best possible for authorship attribution tasks. Their advantage is that they are universal, in the sense that they can be used in any case and provide a robust and accurate model even under difficult scenarios. Any alternative stylometric model, either a new set of features or a combination of several types of features, should be compared with the discussed function word and character n-gram models to prove that it performs better than these baseline approaches.

References

1. Abbasi, A., Chen, H.: Applying authorship analysis to extremist-group web forum messages. IEEE Intell. Syst. **20**(5), 67–75 (2005)
2. Argamon, S., Saric, M., Stein, S.: Style mining of electronic messages for multiple authorship discrimination: first results. In: Proceedings of the 9th ACM SIGKDD, pp. 475–480 (2003)
3. Argamon, S., Whitelaw, C., Chase, P., Hota, S.R., Garg, N., Levitan, S.: Stylistic text classification using functional lexical features. J. Am. Soc. Inf. Sci. Technol. **58**(6), 802–822 (2007)
4. Arun, R., Suresh, V., Madhavan, C.E.V.: Stopword graphs and authorship attribution in text corpora. In: Proceedings of the 3rd IEEE International Conference on Semantic Computing, pp. 192–196 (2009)

5. Benedetto, D., Caglioti, E., Loreto, V.: Language trees and zipping. Phys. Rev. Lett. **88**(4), 048702 (2002)
6. Burrows, J.F.: Not unless you ask nicely: the interpretative nexus between analysis and information. Lit. Linguist. Comput. **7**(2), 91–109 (1992)
7. Chaski, C.E.: Who's at the keyboard?: authorship attribution in digital evidence investigations. Int. J. Digit. Evid. **4**(1), 1–13 (2005)
8. Cristani, M., Roffo, G., Segalin, C., Bazzani, L., Vinciarelli, A., Murino, V.: Conversationally-inspired stylometric features for authorship attribution in instant messaging. In: Proceedings of the 20th ACM International Conference on Multimedia, pp. 1121–1124 (2012)
9. de Vel, O., Anderson, A., Corney, M., Mohay, G.: Mining e-mail content for author identification forensics. SIGMOD Rec. **30**(4), 55–64 (2001)
10. Gamon, M.: Linguistic correlates of style: authorship classification with deep linguistic analysis features. In: Proceedings of the 20th International Conference on Computational Linguistics, pp. 611–617 (2004)
11. Grieve, J.: Quantitative authorship attribution: an evaluation of techniques. Lit. Linguist. Comput. **22**(3), 251–270 (2007)
12. Holmes, D.I.: The evolution of stylometry in humanities scholarship. Lit. Linguist. Comput. **13**(3), 111–117 (1998)
13. Jair Escalante, H., Solorio, T., Montes-y-Gómez, M.: Local histograms of character n-grams for authorship attribution. In: Proceedings of ACL, pp. 288–298 (2011)
14. Jachims, T.: Text categorization with support vector machines: learning with many relevant features. In: Proceedings of the 10th European Conference on Machine Learning, pp. 137–142 (1998)
15. Kanaris, I., Stamatatos, E.: Learning to recognize webpage genres. Inf. Process. Manag. **45**(5), 499–512 (2009)
16. Khmelev, D.V., Teahan, W.J.: A repetition based measure for verification of text collections and for text categorization. In: Proceedings of the 26th ACM SIGIR, pp. 104–110 (2003)
17. Keselj, V., Peng, F., Cercone, N., Thomas, C.: N-gram-based author profiles for authorship attribution. In: Proceedings of the Pacific Association for Computational Linguistics, pp. 255–264 (2003)
18. Koppel, M., Winter, Y.: Determining if two documents are by the same author. J. Am. Soc. Inf. Sci. Technol. **65**(1), 178–187 (2014)
19. Koppel, M., Schler, J., Bonchek-Dokow, E.: Measuring differentiability: unmasking pseudonymous authors. J. Mach. Learn. Res. **8**, 1261–1276 (2007)
20. Koppel, M., Schler, J., Argamon, S.: Authorship attribution in the wild. Lang. Resour. Eval. **45**, 83–94 (2011)
21. Lim, C.S., Lee, K.J., Kim, G.C.: Multiple sets of features for automatic genre classification of web documents. Inf. Process. Manag. **41**(5), 1263–1276 (2005)
22. Luyckx, K., Daelemans, W.: Shallow text analysis and machine learning for authorship attribution. In: Proceedings of the Fifteenth Meeting of Computational Linguistics in the Netherlands (2005)
23. Luyckx, K., Daelemans, W.: Authorship attribution and verification with many authors and limited data. In: Proceedings of the Twenty-Second International Conference on Computational Linguistics, pp. 513–520 (2008)
24. Madigan, D., Genkin, A., Lewis, D., Argamon, S., Fradkin, D., Ye, L.: Author identification on the large scale. In: Proceedings of CSNA-05 (2005)
25. Mendenhall, T.C.: The characteristic curves of composition. Science **IX**, 237–249 (1887)
26. Meyer zu Eissen, S., Stein, B.: Genre classification of web pages: user study and feasibility analysis. In: Biundo, S., Fruhwirth, T., Palm, G. (eds.) KI 2004: Advances in Artificial Intelligence, pp. 256–269. Springer, Berlin (2004)
27. Mosteller, F., Wallace, D.L.: Inference and Disputed Authorship: The Federalist. Addison-Wesley, Reading (1964)
28. Pang, B., Lee, L.: Opinion mining and sentiment analysis. Found. Trends Inf. Retr. **2**(12), 1–135 (2008)

29. Rangel, F., Rosso, P., Koppel, M., Stamatatos, E., Inches, G.: Overview of the author profiling task at PAN 2013. In: Forner, P., Navigli, R., Tufis, D. (eds.) Working Notes Papers of the CLEF 2013 Evaluation Labs (2013)

30. Santini, M.: Automatic identification of genre in webpages. Ph.D. thesis, University of Brighton (2007)

31. Seidman, S.: Authorship verification using the impostors method. In: Forner, P., Navigli, R., Tufis, D. (eds.) CLEF 2013 Evaluation Labs and Workshop Working Notes Papers (2013)

32. Sebastiani, F.: Machine learning in automated text categorization. ACM Comput. Surv. **34**(1), 1–47 (2002)

33. Sidorov, G., Velasquez, F., Stamatatos, E., Gelbukh, A.F., Chanona-Hernández, L.: Syntactic n-grams as machine learning features for natural language processing. Expert Syst. Appl. **41**(3), 853–860 (2014)

34. Stamatatos, E.: A survey of modern authorship attribution methods. J. Am. Soc. Inf. Sci. Technol. **60**, 538–556 (2009)

35. Stamatatos, E.: Intrinsic plagiarism detection using character n-gram profiles. In: Proceedings of the 3rd International Workshop on Uncovering Plagiarism, Authorship, and Social Software Misuse (2009)

36. Stamatatos, E.: Plagiarism detection using stopword n-grams. J. Am. Soc. Inf. Sci. Technol. **62**(12), 2512–2527 (2011)

37. Stamatatos, E., Fakotakis, N., Kokkinakis, G.: Automatic text categorization in terms of genre and author. Comput. Linguist. **26**(4), 471–495 (2000)

38. Stamatatos, E., Daelemans, W., Verhoeven, B., Stein, B., Potthast, M., Juola, P., Sánchez-Pérez, M.A., Barròn-Cedeño, A.: Overview of the author identification task at PAN 2014. CLEF Working Notes, pp. 877–897 (2014)

39. Van Halteren, H.: Author verification by linguistic profiling: an exploration of the parameter space. ACM Trans. Speech Lang. Process. **4**(1), 1–17 (2007)

40. Weiss, S.M., Indurkhya, N., Zhang, T., Damerau, F.: Text Mining: Predictive Methods for Analyzing Unstructured Information. Springer, New York (2005)

41. Yule, G.U.: The Statistical Study of Literary Vocabulary. Cambridge University Press, Cambridge (1944)

42. Zheng, R., Li, J., Chen, H., Huang, Z.: A framework for authorship identification of online messages: writing style features and classification techniques. J. Am. Soc. Inf. Sci. Technol. **57**(3), 378–393 (2006)

Dynamics of Style and the Case of the *Diario Postumo* by Eugenio Montale: A Quantitative Approach

Dario Benedetto and Mirko Degli Esposti

Abstract Here we face a concrete problem of integrity of a very specific textual corpus, namely the 84 poems now forming, after a troubled journey, the so-called *Diario Postumo* (DP) by Eugenio Montale. Our approach is rather simple to describe: it is based on two distinct methods that measure similarity between texts, namely two different algorithms that given any pair of texts return a positive number which is smaller for similar texts. The first similarity measure, called *entropic distance* (or *lzwe*), is based on the use of cross entropy to measure *differences* between sequences of symbols, as learned from data compression theory. The second similarity distance, called *n-gram* distance, is also very simple to describe and it is based on the frequency of sequences of consecutive *n* characters. Both distances have been described elsewhere, but for completeness we report their precise description in the Appendix. The main purpose of our analysis is to test if it is possible, trough purely automatic and quantitative methods, to reveal anomalies in the poems forming the DP. Such anomalies exist and are compatible and coherent with the hypothesis that they are the result of several elaborations of authentic Montale material, originally created and recorded in different forms. Our research on the DP is just a part of a wider research project aimed at exploring the possibility of combining philological qualitative methods with mathematical quantitative approaches to solve problems in authorship attribution (A.A.), forgery detection and integrity texting of textual corpora.

1 Introduction: The Problem, Approach, and Methods

A typical scenario: after my seminar, a colleague from human sciences takes me apart and asks

D. Benedetto (✉)
Dipartimento di Matematica, Sapienza Università di Roma, Rome, Italy
e-mail: benedetto@mat.uniroma1.it

M. Degli Esposti
Dipartimento di Matematica, Università di Bologna, Bologna, Italy
e-mail: mirko.degliesposti@unibo.it

© Springer International Publishing Switzerland 2016
M. Degli Esposti et al. (eds.), *Creativity and Universality in Language*,
Lecture Notes in Morphogenesis, DOI 10.1007/978-3-319-24403-7_10

...but why should a philologist and two mathematicians work together? Do the second ones really help the first to determine the author of a given text? And does this approach also work in the case of the *Diario Postumo* by Montale? What does the mathematician exactly do? Does he write equations? Let me see, come on, show me, now!!

Now a small premise is necessary. Analyzing and modeling textual data, or more generally understanding and measuring the deep mechanisms underlying the evolution and organization of language (in all its multiple forms) constitute an incredible source for new mathematical models and novel computational techniques. Other contributions in this volume confirm the variegated and solid research in mathematical physics stimulated by languages and textual data. In this paper we do something else. We report on a recent experience, where we joined experts with radically different approaches to attack a concrete problem: the integrity of a very specific textual corpus, the 84 poems now forming, after a troubled journey, the so-called *Diario Postumo* (DP, from now on) by Montale.[1] Montale died in 1981, the first six poems of the DP appeared for the first time only in 1986. They were contained in an envelope (*busta* in italian) brought to the public by Annalisa Cima, a young woman that since around 1968 had established a well-documented intellectual relationship with the poet.[2] Since 1986 and until 1995 this process continued and each year a new envelope was revealed by Annalisa Cima, always containing six new poems, until 1996 when (as a surprise) a bigger envelope (*bustone* in Italian) with 24 original poems which made up the 84 poems of the DP. Annalisa Cima always asserted that this was a "time-bomb" carried out following Montale's own precise instructions and it is quite surprising to note the coincidence between the completeness of the DP with the 100 years anniversary of Montale. The DP immediately caused a scandal in Italian literary circles. Some critics believed that the poems were composed by Cima out of conversations with Montale, while others believed Cima had forged them outright. Then after a short period of vivid discussions, everything cooled down to a period of silence, the DP was officially included in Montale's bibliography and published worldwide. Then very recently Federico Condello reopened the discussion with a deep investigation of the events that brought to the surface new facts and contradictions, adding more shadows on the integrity of the DP [8]. Since Eugenio Montale is one of the highest representative of Italian literature, the recent and renewed discussion about the integrity of the DP-fuelled fervent dispute among humanists, at least in the Italian community.

When we turn our methods of analysis and our mathematical and computational techniques to a such specific scenario, they reveal, as we will see, indications sup-

---

[1]Eugenio Montale (12 October 1896–12 September 1981) was an Italian poet, prose writer, editor and translator, and recipient of the 1975 Nobel Prize in Literature. He is widely considered as the greatest Italian lyric poet since Giacomo Leopardi (https://en.wikipedia.org/wiki/Eugenio_Montale).

[2]Annalisa Cima is born in 1941. Her father was leading a very wealthy paper industry (*Industrie Cartarie Cima*), while her mother was related to the Schlesinger, an Austrian aristocratic family. When Miss Cima met Montale for the fist time, she was 27, with interests in painting and poetry, well accustomed to establish the worldwide intellectual relationships with various (often older) artists of the period [8].

porting the concrete doubts about the integrity of the DP and coherent with the hypothesis of external manipulations of an initial original material by Eugenio Montale. Concrete doubts recently revealed and supported by a deep and documented historical and philological investigation (see below). It should not come as a surprise that in facing this problem we will be rigorous, but we will not reach rigorous definitive conclusions. Nevertheless, we are also convinced that these kind of research might stimulate novel mathematical problems, as we will argue in the section about measuring the evolution of Montale's writings.

With this in mind, we can come back and try to answer our colleague from human sciences: no, a mathematician in this case does not write equations and in the case of the attribution of the *Diario Postumo* things are rather complicated (and for this reason more interesting). First of all, we will not have *equations* to help us, as an equation is either satisfied or not, either it is verified for certain values of the unknowns or not, there is no room for uncertainty. Exactly, the opposite situation that we face in an *Authorship Attribution* (A.A.), when we try to identify the author (or several authors) of a given text trough quantitative methods. Any conclusions we can reach about a possible author are subjected to all conceptual and practical limits that come with probabilistic–statistical methods.

Nevertheless, the mathematical history of A.A. is not recent and we can identify is first steps with the work of the American physicist and meteorologist T.C. Mendenhall (1841–1924). In 1887, attracted by the dispute about the real identity of William Shakespeare, Mendenhall tried to answer to this question by considering the works of Christopher Marlowe and by calculating the so-called *words length spectrum* [15], essentially the frequency histogram of words with a given character length (for a deeper introduction to the history of A.A. see [10, 17]). Although we now know very well that this spectrum is by far not able to characterize an author, Mendenhall's tentative was the beginning of a quantitative approach to attribution problems. Nowadays, several quantitative approaches to A.A. have been developed, all featuring two aspects, both of which play a crucial role in our approach to the DP.

The first aspect is related to the nature of the *objects* that are extracted from the text for measurement. In the last three decades, mostly due to the important influence of information theory, the attention of researchers shifted from the analysis of *natural* objects, such as single words or other syntactic elements, to more abstract objects, such as the so-called *n-grams*, i.e., consecutive sequences of *n* characters, where a character is any symbol, space included, that appears in the text [11, 17] Furthermore, the techniques used for statistical analysis have been extended from the simple frequencies of textual objects to their returning time and to the exploration of their (long-range) correlations [1].

The second important aspect concerns the techniques that are used to combine the information extracted from the texts and to transform it in assertions about the final attribution. One of the main approaches of quantitative A.A. comes from the general theory of machine learning that reduces any A.A. problem to a classical problem of "automatic classification based on feature extraction."

In few words, at least for the colleagues in human sciences, one extracts specific features (phonetic, grammatical, syntactic, semantic, and statistical) from texts, thus representing each text by a vector, usually of very high dimension, where each component represents a value of a given measured property. Then automatic algorithms able to learn from the data are trained on data with known authors. Given data from a candidate text, they return some prediction about the attribution, namely the author more compatible/similar with the vector representing the candidate text (see [10, 17] and also the contribution by Efstathios Stamatatos in this Volume for a more precise discussion). Our approach is based on two distinct methods that measure similarity between texts, using two different algorithms that given any pair of texts return a positive number which is smaller for similar texts, hoping that the captured similarity is only due to the author's style and not to other kind of similarity, for example the semantic similarity in case of texts discussing the same topic.

The first similarity measure is based on the use of cross entropy to measure *differences* between sequences of symbols, as learned from complexity theory and from the related information theory [9, 19]. Certain algorithms with practical use in computer science, such as the ones used in data compression, offer techniques that can be implemented to obtain *good* approximations of the cross entropy [3]. The specific algorithm that we use here, called *lzwe* and inherited from compression algorithms, automatically detects the existence of common substrings of characters between two given texts, weighting their frequency, length, and position among both texts, and calculating, following the prescriptions of information theory, what we will call an *entropic distance* (or also distance *lzwe*) between the two texts [4]. The second similarity distance is also very simple to describe and will be called *n-gram* distance: fix a given integer value n (typically and also for the DP, $n = 6$), then for each text just compute the frequency of each characters n-gram appearing at least once (i.e., any sequence of consecutive n characters, space included); then one simply defines a distance based upon these frequencies by taking care of weighting less the difference for very frequent n-grams [2]. In this way, typical but sporadic n-grams for an author will give a substantial contribution to the distance. Both distances have been described elsewhere but for completeness, we report their precise description in the Appendix. Concerning the problem we face here, it is now sufficient to remark that the specific differences among the two measures allow us to state that while *lzwe* capture similarities in the *background style*, hidden in the statistics of frequent but short sequences, the 6-gram distance reveals and amplifies rare but recurrent characteristics at the scale of words. This is a very simple but essential observation that will play a relevant role in our conclusions concerning the DP. Our attribution procedure simply relies on calculating distances between an unknown text and other texts of known author. Because these intertextual distances strongly depend on the lengths of the texts, we first compare the unknown text with chunks of fixed length. These procedure produces a set of positive numbers, i.e., the similarity measures between the given unknown fragment and all chunks of known attribution. These numbers are then turned into an attribution by a weighted voting system that elaborates them in a quite simple way, returning the value of an *author index* that measure the possible attribution to a given author [2, 4]. Mathematical details about our weighted voting system can also be found in the Appendix, where calculations for the case of two authors are reported from [4] (in the DP case, for reasons we will discuss, we will use the voting algorithm with three authors). The efficacy of this class of methods in solving A.A. problems (but also in plagiarism detection) has been widely verified and discussed [17, 18]. Our interest in the DP case is to verify if in a concrete case of attribution (as we will see very far from the artificial scenario traditionally faced), a suitable use of these methods could effectively give coherent and scientifically correct, even if not conclusive, indications. Our research on the DP is just a part of a wider research project aimed at exploring the possibility of combining philological methods with mathematical quantitative approaches to solve problems in A.A., forgery detection, and integrity of textual corpora. The main purpose of the analysis that we will discuss here is hence to understand if it is possible, trough purely automatic and quantitative methods, to reveal anomalies in the poems forming the DP.

As we will see, such anomalies exist and are compatible and coherent with the hypothesis that they are the result of several elaborations of authentic Montale material, originally created and recorded in various forms (textual data, audio, notes, sketches, etc...). The solidity of this hypothesis relies, in our opinion, upon the conclusions of the work presented and discussed by Federico Condello in his recent book (in Italian) *"I filologi e gli angeli. E' di Eugenio Montale il Diario postumo?"* ([8], see also [16]). Condello's findings are based on techniques completely different from ours, but we had the privilege to follow this work from the beginning. It started with a deep and complete historical and philological reconstruction of all the facts and contradictions that characterize the genesis and the transmission of the individual poems forming the DP. This first decisive study revealed numerous and very solid doubts about the integrity of the DP. It was subsequently supported by two different studies that explored calligraphic features of the *original* writings.[3] The first calligraphic study raised serious doubts on the authenticity of Montale's signature, while the second showed that the hand writing was not compatible with the well-known and coeval degenerative neurological state of the poet [8]. Furthermore, additional indications come also from a quasi-quantitative analysis performed on the metric structure of the poems [16].

Well, maybe our the colleague from humanities won't be very satisfied after reading our results, but this is very understandable. This case is too peculiar and too complex to expect precise answers from quantitative methods, but nevertheless we believe it is important to continue fertilizing interdisciplinary approaches on concrete cases of authorship attribution and integrity of textual data.

2 *Diario Postumo*: An Anomalous A.A. Problem and the Small Scales

As soon as we attack the question about the authenticity or integrity of each one of the 84 poems contained in the DP as they arrived to us after the years long trouble adventures of *buste* (envelopes) and *bustoni* (large envelops), we immediately realize

---

[3]To our knowledge, the original writings are not directly accessible and only photographic copies of them are available.

as this problem is really far from any traditional A.A. problem. The DP case has at least three very peculiar characteristics that make it unique and ask for a very specific approach.

1. We are facing a textual corpus (84 poems) that for sure contains original fragments from Eugenio Montale, even if probably some of them were turned into textual material only later and not directly by the Poet. We are thinking about pieces of oral discussions, scattered notes or audio recordings that, as we know now vey well, characterized for years the relationship between the poet and his self-proclaimed muse [8]. It is then not possible to consider the DP as a simple problem of authorship attribution. On the contrary, one has to explore the possibility of measuring the integrity of the corpus, eventually emphasizing possible interventions that might have modified the original content.
2. Texts are very short and poetic. It is undeniable that all quantitative methods ultimately rely on the statistics of the extracted features and that the efficiency of the methods strongly increase in the presence of long texts (say around 10 k characters). Very rarely have quantitative methods alone[4] been used to attribute very short texts (few thousands or few hundreds of characters). But undoubtedly the modern methods of text generation (blog, social networks, emails,...) strongly require novel methods for A.A. and plagiarism detection able to work efficiently at these very small scales.
3. Poems of the DP have a chronological order: 82 poems out of the 84 are assigned each to a precise year inside the decade 1969–1979. This chronological information is crucial for a precise reconstruction of the genesis of the corpus and plays a fundamental role in accounting for all the coincidences and contradictions that characterize the history of the DP [8].

3   Texts and Hypothesis

Here a description of the texts will be used for our analysis. As a first step, we have created a corpus containing the known works by Eugenio Montale and Annalisa Cima (see Tables 1 and 2).

For reasons we will explain later that we have used two different prefixes "ma" and "mb" for labeling the works by Montale that have been either created until or after the collection of poems called *"La bufera"*. This subdivision of Montale's corpus in two subsets "ma" and "mb" is suggested by well-known stylistic and philological considerations that identify a sharp discontinuity point after *"La bufera"* in the style of the poet.[5]

---

[4]I.e., methods that do not make direct use of either linguistic or semantic informations.
[5]Suggested by Federico Condello.

Table 1 Works by Eugenio Montale used in the analysis

Author: Eugenio Montale

Opera	Label	Year	N. of characters
Ossi di seppia	ma1	1920–1927	51058
Le occasioni	ma2	1928–1939	38838
La bufera e altro	ma3	1940–1954	42626
Satura	mb4	1962–1970	61239
Diario del 71 e del 72	mb5	1971–1972	49847
Quaderno dei 4 anni	mb6	1977	60553

Table 2 Works by Annalisa Cima used in the analysis; in order to have an homogeneous distribution of lengths, we have collected together texts consecutive in time, forming the texts $c1$, $c2$, and $c3$ as shown

Author: Annalisa Cima

Opera	Label	Year	N. of characters
Terzo mondo	c00	1969	5392
La Genesi e altre poesie	c01	1971	15368
Immobilità	c02	1974	6801
	c1		**28101**
Sesamon	c03	1977	14400
Ipotesi d'Amore	c04	1984	7288
Aegri somnia	c05	1989	3649
	c2		**25337**
Quattro canti	c06	1993	2564
Il tempo predatore	c07	1997	7712
Hai ripiegato l'ultima pagina	c08	2000	6335
Canto della primavera e della sopravvivenza	c09	2001	8398
Poesie Pulcino Elefante	c10	2001	2181
	c3		**27190**

A first, simple but essential numerical experiment consists of verifying if and how our methods are capable to distinguish between these two authors at the scale of a single work. In order to verify this, we simply take out each single work and then attribute it using the other works as reference set. Namely, given one of the two methods, we calculate the distances between a given work and all the others. We then turn these distances into an attribution using the weighted voting systems (recalled in the Appendix).

It is welcome, but not surprising, to see that both methods (*lzwe* and 6-gram) reach a clear 100 % of correct attributions.

It is also remarkable that attribution remains very high even if one tries to attribute shorter textual fragments of few thousands characters (data not shown). But as we will see, this is highly nonsufficient and we need to get down to much shorter scales. We will come back to this and now let us look at the evolution of style.

4 Before and After: Measuring the Evolution of Style

The next analysis consists of verifying if our methods return results that are consistent with the subdivision of Montale's corpus in two subset, "ma" and "mb", as suggested by stylistic and philological considerations. As we will now see, these results are quite positive and stimulate future developments: in our opinion, quantitative studies aimed at measuring and modeling the evolution of literary styles deserve more attention and the future research is desirable.

Given any one of the two methods (*lzwe* or 6-gram), we calculate the distance matrix between any two pairs of segments obtained by cutting each work of Montale into nonoverlapping fragments of size approximately 4 000 characters. Then by averaging the distance among fragments belonging to the same work, we build the matrix that contains the averaged distances between single works. Finally, we build the phylogenetic tree of the works by Montale using the so-called Neighbouring Joining (NJ) algorithm, commonly used in evolutionary biology. Essentially, the algorithm finds (one of) the best tree with leafs identified with Montale's works, that is compatible with the measured mean distances: the distance between two given texts (leafs of the tree) is given by summing the lengths of the branches joining the two leafs (because the graph is in fact a tree, this path joining the two leafs is unique) and it is a good approximation of the averaged distance measured using one of our similarity distances. In Fig. 1 we show just the topology of the tree: in this representation, all branches are drawn with the same length. The result is perfectly compatible with the real chronological order of Montale's works and also compatible with the hypothesis of a sharp discontinuity in style after *"La bufera"* (ma3).

In Fig. 2, we show instead the topology of the tree obtained by also inserting texts c1, c2, and c3 by Annalisa Cima. Note that Cima's works are grouped in a very distinct branch, located in the middle of the two distinct periods of Montale's production.

It could be interesting to use this tree as a tool to analyze the poems of DP. In particular, we should find the nodes from which to derive the individual poems as additional leaves, and use the location of these nodes on the tree for attribution; work is now in progress in this direction.

Fig. 1 Topology of the Montale's works tree. The tree is obtained by averaging the distances between fragments of about 4000 characters. Both entropic and 6-gram distance give the same tree

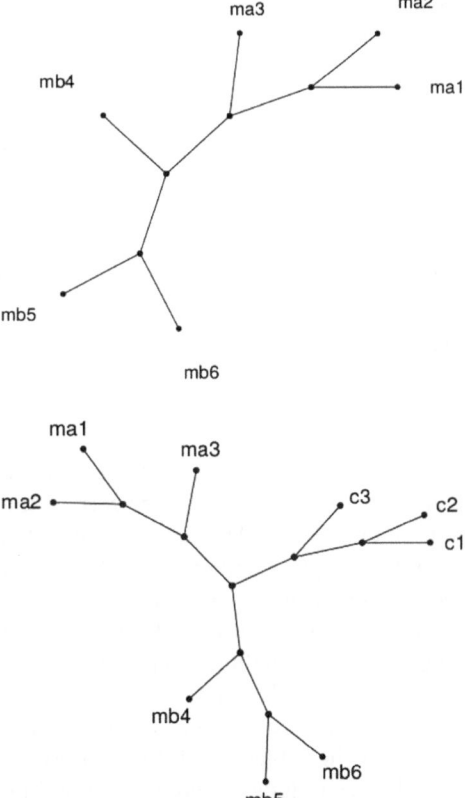

Fig. 2 The tree obtained from both Montale and Cima works. The tree is obtained by averaging the distances between fragments of about 4000 characters. Both entropic and 6-gram distance give the same tree

5 Going to Smaller Scales

In the previous analysis, we compared segments of about 4000 characters, a quite large scale; in order to reach smaller scales, e.g., the ones typical of a single poem of the DP, and still preserving a robust procedure we need to smooth out the fluctuations induced by the nonhomogenous lengths distribution of the reference corpus. For this we have decided to segment the reference texts in bigger segments of about 12000 characters each, both for Montale and Cima (see Table 3).

As a preliminary test, we have verified the efficiency of our methods at small scales by trying to attribute all single poems with known authorship (keeping the distinction ma, mb, i.e., a three authors problem "ma", "mb", "c"). In Table 4 we show the number of distinct poems in each work, while in Table 5 we show the distribution of the lengths of individual poems in the three reference sets (ma, mb, c).

Various numerical experiments confirmed the stability of the attribution methods also at such small scales. For example in Table 6, we report the attribution of frag-

Table 3 Number of fragments of about 12 000 characters in Montale and Cima corpus

Label of the text	Number of fragments
c1	2
c2	2
c3	2
ma1	4
ma2	3
ma3	3
mb4	5
mb5	4
mb6	5

ments of 400 consecutive characters. More precisely, we consider a segment of 400 consecutive characters from the corpus (ma, mb, c) and we calculate its *lzwe* distance to each one of the 30 bigger segments of about 12 000 characters obtained from the reference corpus (Table 3), where of course the considered small segment but also the two segments of 400 characters before and after it have been deleted. These 30 distances are then turned into an attribution by the weighted voting procedure adapted to three authors "ma", "mb", "c" (see Appendix for the calculations in the (similar) case of 2 authors).

Similarly, in Table 7 we present the results obtained using the 6-gram procedure.

Finally, in Tables 8 and 9 we report the attribution results for segments of only 200 characters long, where again the 400 characters long segments following and

Table 4 Distribution of the number of poems in the works of Eugenio Montale and Annalisa Cima

Eugenio Montale		Annalisa Cima	
Work	Number of poems	Work	Number of poems
ma1	59	c00	12
ma2	58	c01	7
ma3	58	c02	73
mb4	94	c04	36
mb5	86	c05	13
mb6	115	c06	4
		c07	4
		c08	50
		c09	3
		c10	11

Table 5 Statistics of the lengths of individual poems

Length	cima	ma	mb	DP
49–100	73	0	4	0
100–200	92	2	22	4
200–400	67	57	96	49
400–800	18	61	111	30
800–1 600	8	36	56	1
1 600–3 200	6	18	5	0
3 200–6 400	2	1	1	0

Table 6 Attributions of fragments of about 400 characters

lzwe				Attributions
Author	c	ma	mb	
c	175	17	6	175/198 = 88.38 %
ma	0	314	16	314/330 = 95.15 %
mb	0	35	393	393/428 = 91.82 %
				total 882/956 = 92.26 %

lzwe method

Table 7 Attributions of fragments of about 400 characters

6-gram				Attributions
Author	c	ma	mb	
c	193	3	2	193/198 = 97.47 %
ma	6	288	36	288/330 = 87.27 %
mb	4	5	419	419/428 = 97.90 %
				total 900/956 = 94.14 %

6-gram method

preceding the analyzed segment have been deleted from the reference corpus (then the procedure remains the same as before).

It is interesting to notice that also at very small scales, like segments of only 200 characters, the entropic method *lzwe* reaches quite surprising levels of correct attributions. For example, if we concentrate only on the discrimination Cima/Montale, we reach almost 95 % of positive attributions (Table 10).

Table 8 Attributions of fragments of about 200 characters

lzwe				Attributions
Author	c	ma	mb	
c	335	48	15	335/398 = 84.17%
ma	3	618	41	618/662 = 93.35%
mb	10	114	733	733/857 = 85.53%
				total 1 706/1 917 = 88.99 %

lzwe method

Table 9 Attribution of fragments of about 200 characters

6-grammi				Attributions
Author	c	ma	mb	
c	378	10	10	378/398 = 94.97%
ma	36	522	104	522/662 = 78.85%
mb	20	27	810	810/857 = 94.52%
				total 1 740/1 917 = 90.77 %

6-gram method

Table 10 Attribution Montale/Cima of segments of about 200 characters—*lzwe* method

lzwe			Attributions
Author	c	m	
c	378	20	378/398 = 94.97%
m	80	1 439	1 439/1 519 = 94.73%
			total 1 817/1 917 = 94.78 %

Moreover, we can also appreciate how the data reveal certain stability and coherence of the two methods, both producing comparable and compatible results.

6 The DP: Discussion and Conclusions

We can now turn to the attribution of each single poems of the DP, again using both the *lzwe* and 6-gram methods, always comparing each single poem with the (nonoverlapping) segments of about 12 000 characters from the reference corpus of each author and again turning distances into attribution to one of the three authors "ma", "mb", "c".

In the tables below we show, for each poem, the attribution results as obtained by both methods. The first value indicates the length (in characters) of the poem (the shorter the length, the less reliable the attribution). Then, for each attribution method, we show the authorship indexes as calculated by applying the weighting method to the measured distances relatively to Cima ("c"), Montale before *La bufera* ("ma") and Montale after *La bufera* ("mb"). The attributed author is the one with a greater index value. In the last two columns, we show the corresponding attributions.

		c	ma	mb	c	ma	mb	lzwe	6-gram				c	ma	mb	c	ma	mb	lzwe	6-gram
d01	208	-3.82	-2.27	6.09	-2.19	-6.37	8.56	mb	mb		d43	330	0.75	-3.57	2.82	3.27	-5.94	2.68	mb	c
d02	345	0.23	3.96	-4.19	0.84	6.14	-6.97	ma	ma		d44	350	-0.48	-2.33	2.81	6.87	-3.06	-3.82	mb	c
d03	267	2.04	2.96	-5.00	3.74	0.18	-3.93	ma	c		d45	391	4.04	-7.80	3.76	6.77	-7.48	0.71	c	c
d04	455	-2.80	0.71	2.09	3.07	-3.62	0.55	mb	c		d46	237	-0.53	8.78	-8.24	-0.59	8.23	-7.64	ma	ma
d05	217	1.52	-2.44	0.93	0.09	-2.03	1.94	c	mb		d47	420	-0.58	-5.74	6.32	2.53	-7.25	4.72	mb	mb
d06	488	-2.52	3.77	-1.25	2.14	0.61	-2.75	ma	c		d48	350	2.22	-1.30	-0.92	5.60	-3.46	-2.14	c	c
d07	368	0.43	2.26	-2.69	0.55	1.38	-1.93	ma	ma		d49	345	-1.59	-1.40	2.99	4.33	0.97	-5.30	mb	c
d08	300	-0.91	9.39	-8.48	1.96	5.35	-7.31	ma	ma		d50	950	0.97	2.09	-3.07	5.04	-3.09	-1.96	ma	c
d09	184	3.33	-5.52	2.20	2.99	-6.60	3.61	c	mb		d51	486	3.33	-3.91	0.58	2.68	-5.09	2.41	c	c
d10	480	0.42	-4.12	3.70	1.63	-5.29	3.66	mb	mb		d52	286	-1.26	0.69	0.58	1.58	-2.65	1.07	ma	c
d11	415	-4.26	3.35	0.91	-1.14	-5.20	6.34	ma	mb		d53	317	-2.31	8.10	-5.78	-0.75	2.51	-1.76	ma	ma
d12	563	-4.31	-2.43	6.74	-1.91	-6.51	8.42	mb	mb		d54	330	6.38	-0.44	-5.95	5.05	-6.28	1.23	c	c
d13	179	3.67	-7.15	3.48	2.64	-7.28	4.65	c	mb		d55	468	2.21	-0.57	-1.64	7.54	-2.40	-5.15	c	c
d14	205	0.12	1.20	-1.31	3.07	-4.49	1.42	ma	c		d56	345	0.13	-1.39	1.27	4.30	-4.25	-0.05	mb	c
d15	254	2.15	-1.33	-0.82	7.37	-5.34	-2.03	c	c		d57	683	1.93	3.77	-5.70	3.05	-1.93	-1.12	ma	c
d16	234	-2.98	8.31	-5.33	1.45	1.34	-2.79	ma	c		d58	619	-0.44	-7.06	7.49	1.73	-7.94	6.21	mb	mb
d17	652	-2.63	9.64	-7.01	-1.44	8.24	-6.80	ma	ma		d59	371	-3.58	-2.37	5.95	-0.79	-6.52	7.31	mb	mb
d18	356	1.88	-0.67	-1.21	3.73	-3.41	-0.32	c	c		d60	268	-0.11	-5.95	6.06	3.54	-7.22	3.67	mb	mb
d19	291	-3.07	-1.23	4.30	-2.78	-4.86	7.64	mb	mb		d61	489	1.10	4.53	-5.63	4.80	-1.76	-3.04	ma	c
d20	401	-2.51	-4.17	6.68	-3.29	-6.61	9.90	mb	mb		d62	443	-3.56	-4.53	8.09	-2.36	-6.82	9.19	mb	mb
d21	269	0.69	-5.29	4.60	1.36	-7.28	5.92	mb	mb		d63	460	-4.89	1.23	3.66	-2.61	-1.21	3.81	mb	mb
d22	595	0.03	-7.11	7.08	2.64	-7.84	5.20	mb	mb		d64	607	-3.22	4.24	-1.02	5.66	-3.74	-1.92	ma	ma
d23	395	-2.78	-5.27	8.05	-1.12	-6.13	7.25	mb	mb		d65	373	3.03	3.57	-6.60	5.74	2.74	-8.49	ma	c
d24	460	-1.01	6.52	-5.51	2.02	2.94	-4.96	ma	ma		d66	405	-1.93	-5.97	7.90	-2.13	-7.22	9.36	mb	mb
d25	307	1.24	0.06	-1.30	3.48	-5.17	1.69	c	c		d67	412	0.23	8.01	-8.24	1.53	2.03	-3.56	ma	ma
d26	258	3.39	4.19	-7.58	5.69	-1.34	-4.35	ma	c		d68	738	0.95	5.79	-6.73	7.16	-0.94	-6.42	ma	c
d27	253	1.23	5.40	-6.64	5.16	-0.80	-4.36	ma	c		d69	315	-1.36	1.29	0.06	2.97	-6.05	3.08	mb	mb
d28	198	-0.03	-7.46	7.49	2.61	-7.89	5.29	mb	mb		d70	262	-1.73	5.40	-3.68	0.99	4.49	-5.48	ma	ma
d29	349	-3.67	3.49	0.18	0.73	-1.19	0.46	ma	c		d71	299	0.27	7.42	-7.69	3.67	3.89	-7.56	ma	ma
d30	526	-4.17	6.75	-2.57	2.65	1.29	-3.94	ma	c		d72	121	-4.81	-0.60	5.40	-1.98	4.67	-2.68	mb	ma
d31	240	-1.88	-2.34	4.22	-3.24	-4.11	7.35	mb	mb		d73	439	-3.93	-3.36	7.29	1.59	-6.94	5.35	mb	mb
d32	371	-2.77	-6.25	9.02	-1.22	-7.80	9.02	mb	mb		d74	184	-4.29	7.84	-3.55	-1.62	6.10	-4.48	ma	ma
d33	358	-2.68	-0.57	3.25	-3.17	-3.02	6.20	mb	mb		d75	278	-1.17	2.11	-0.94	-2.99	1.95	1.04	ma	ma
d34	390	-3.74	7.74	-4.00	3.47	0.14	-3.61	ma	c		d76	394	-1.84	0.83	1.00	6.03	-5.32	-0.71	mb	c
d35	389	-2.62	2.78	-0.15	3.84	-0.35	-3.49	ma	c		d77	425	2.90	2.30	-5.21	4.85	-1.34	-3.52	c	c
d36	392	2.43	-5.13	2.70	8.37	-7.10	-1.27	mb	c		d78	265	0.45	0.17	-0.62	3.61	-2.98	-0.63	c	c
d37	405	-0.67	3.39	-2.72	6.18	-6.47	0.29	ma	c		d79	319	1.52	-7.09	5.57	4.76	-7.80	3.04	mb	c
d38	341	0.26	-0.60	0.34	3.85	-5.45	1.59	mb	c		d80	408	-0.40	7.50	-7.10	3.50	0.44	-3.94	ma	c
d39	354	-2.09	-5.11	7.20	1.73	-3.88	2.15	mb	mb		d81	391	-1.85	7.34	-5.48	-0.65	-0.46	1.11	ma	mb
d40	253	0.12	1.89	-2.01	2.69	-0.91	-1.78	ma	c		d82	688	1.06	2.79	-3.86	6.23	1.23	-7.46	ma	c
d41	480	-1.34	-2.88	4.22	5.25	-6.25	1.00	mb	c		d83	318	-0.72	3.28	-2.56	2.40	0.09	-2.50	ma	c
d42	246	-1.06	3.27	-2.21	-0.07	0.02	0.05	ma	mb		d84	244	0.60	-6.55	5.95	0.87	-6.79	5.93	mb	mb

We shall now concentrate on what, given the analysis and the data, can be considered as the main conclusion: the entropic method *lzwe* attributes almost 11 % of the DP's poems (13 out of 84) to Cima, while this percentage jumps to 50 % (42 out of 84) when distances are calculated using the 6-gram method. The discrepancies between the attributions by the two methods when applied to the DP, compared with the robust indications obtained when applied to texts with known single authorship, reveal strong anomalies in the DP. Therefore, at least from the point of view of quantitative methods, the DP cannot be considered as a genuine and uncorrupted work by Eugenio Montale. Based upon the differences between the two methods, as previously discussed, we can formulate an hypothesis: *lzwe* enhances frequent similarities at smaller scales, while the 6-gram distance amplifies differences in words usage and in other quite rare structures. The data strongly support a scenario where precise interventions in a style that can be traced back to Cima have been performed on an existing textual material written in a style consistently attributed to Montale. We believe that our analysis reveals new shadows about the process that generated the DP as it is known now (see again [8] for more details).

It is important to stress that, due to the dimension of individual poems, our conclusions are not supported by solid statistical data and that our results alone do not allow any serious conclusion concerning the attribution of the pomes of the Diario Postumo. For this reason also the comparison between our work and a first quantitative approach to the DP as presented by Paola Italia and Paolo Canettieri [6] should be considered with extreme caution.

We conclude by mentioning that there exists an interesting version of the classic A.A. problem, namely the so-called *Authorship Verification* (A.V.): one has to verify if a given text has either been written by a given author or not. This one-class problem, with only one author eligible for the attribution, is much more difficult than A.A., where one chooses a possible author among a certain number of candidates. This is because our distances, used to detect similarities, do not have a *natural scale*. One cannot tell without comparison, if a given distance value is either big or small, in other words if it does point out differences or similarities. Nevertheless, there are some recent developments in A.V. [18] that could be investigated and adapted to the DP case.

Appendix

For the sake of completeness, we review here in detail the two similarity distances we have used in the analysis, already described in [2, 4]. As already stressed, a simple but crucial assumption underlying all of our mathematical techniques is that a text should be thought "just" as a sequence of symbols chosen from an alphabet, not taking into consideration either the semantic content of the text or its linguistic/syntactic/grammatical aspects: letters of the alphabet, punctuation marks, blank spaces are just abstract symbols, without any hierarchy between them.

As a consequence, the word as basic constituent of the text has no more meaning than any other aggregate of symbols, and its role as higher level unit is taken up by the (character) n-gram.

6.1   N-Gram Distance

Here we recall and essentially plagiarize parts of our papers [2, 4], for which we refer more details and references: we call ω an arbitrary n-gram, and we denote by $f_X(\omega)$ ($f_Y(\omega)$) the relative frequency with which ω occurs in text X (Y). $D_n(X)$ is the *n-gram dictionary* of text X, that is, the set of all n-grams which have nonzero frequency in X (similarly for Y) and we define what we will call the n-gram distance between text X and text Y as:

$$d_n(X, Y) := \frac{1}{|D_n(X)| + |D_n(Y)|} \sum_{\omega \in D_n(X) \cup D_n(Y)} \left(\frac{f_X(\omega) - f_Y(\omega)}{f_X(\omega) + f_Y(\omega)} \right)^2. \qquad (1)$$

Here, $|D_n(X)|$ and $|D_n(Y)|$ are the numbers of different n-grams in the two dictionaries, respectively, and the sum is taken over all the different n-grams occuring in the two texts.[6] Even if this distance is probably one of the simplest possible measures between texts, it has a relatively short history in published bibliography: after a first experiment based on 2-gram frequencies presented in 1976 by W.R. Bennett [5], V. Kešelj et al. [11] published in 2003 a paper in which n-grams frequencies were used to define a similarity distance between texts (see also [7]).

6.2   lzwe

Again reporting for completeness, the description and the tables already contained in [2]. The *lzwe* distance is based on *data compression*, nowadays there is a very well-established field of information theory, thanks to the founding papers published by J. Ziv, A. Lempel and their coworkers in the 1970s (cf., among others, [12–14] and the review paper [20]). They proposed a variety of compression algorithms (the family of) *LZ algorithms*, based on the idea of a clever *parsing* (subdivision) of the symbolic sequence, i.e., to split it into pieces so that this separation can then be used to produce a shorter, equivalent version of the string itself.

In 1993 J. Ziv and N. Merhav [21] proposed a method to estimate the relative entropy (or Kullback–Leibler divergence) between couples of information sources. This important result was used in various subsequent studies, among which [3] and [2], to deal with problems of text classification and clustering.

Following the idea of LZ77 algorithm, we estimate the similarity of the two sequences in the following way:

---

[6]Note that in contrast with what happens for the Euclidian distance, each term of the sum is weighted with the inverse of the square of the sum of the frequencies of that particular n-gram. In this way *rare words*, i.e., n-grams with lower frequencies, give a larger contribution to the sum.

1. we sequentially parse x starting from the first character in such a way that each string of the parsing is the longest possible sequence that occur at least once in y;
2. the length of each substring and its position (*index*) in y (starting from the end of y) are stored;
3. the coding data, given by the lengths and the positions of the subsequences, are then sequentially compressed.

For example, consider the two sequences

$$y = aaabbababbbaaba \text{ and } x = ababaabbbaa$$

The text x is parsed into the substrings *abab aabb baa*, of lengths 4, 4, 3 and indexes 10, 14, 5 respectively; then these numbers are suitable codified. It should be clear that the longer are the common sequences, the shorter are the numbers of bits we need to reconstruct x given y.

In gzip a match shorter than 3 characters is always ignored and moreover it can never have a length larger than 258 characters. Characters not belonging to an accepted match are directly memorized with a corresponding 8 bits coding, whereas lengths larger than 2 are coded in groups with integers between 257 and 271, using some extra bits to distinguish between lengths with the same code. In the same way, positions of the matches between 1 and 2^{15} are coded as by gzip using integer numbers between 0 and 29 (the codex of the position) and some more extra bits.

Finally, length codes (together with characters with no match) and index codes are independently compressed using a Huffman coding. This is basically the entropic algorithms (called BCL) already used in [2] for the attribution of Gramsci's articles. As we can see from the description, there is a limitation due to the coding of the indexes.

In concrete scenarios one face the attribution of short fragments of text, sometimes against large reference corpus, even larger than 2^{15}, the gzip limitation. These facts require a more sensible parsing procedure (i.e., faster converging relative entropy approximation) and a new coding of the indexes.

For this reason, again inspired by gzip and some if its variants (such as *LZMA*), we have modified the parsing procedure and the coding of the indexes. In particular, the strings forming the parsing of y have both the first and last character in common. Namely, each new match begins with the last character of the previous match.

Let σ be this first character of a new match. As before, we have to store the position of the match in y. Instead of storing the number of characters from the match and the end of y, we store as index of the position the number of words in y that starts with σ and lying between the match and the end of y. For example, consider again the sequences x and y above. Our algorithm first returns the character a, then the position of the match *abab* is indicated by the value 5, just because *abab* is the substring starting with the fifth a from the end; then the match *baab* is found (starting from the final b of the previous match), and its index is 2. Finally, the last match, *bbbaa*, has index 4.

Table 11 For each literary work, we show the author (left), the dimension (dim) in characters and the compression rate (i.e., bytes/character) using `gzip` and `lzwe`, respectively

Author	Title	dim	gzip	lzwe
G. Galilei	La bilancetta	8 936	3.04	2.85
N. Machiavelli	Favola di Belfagor arcidiavolo	19 625	3.30	3.11
D. Alighieri	Quaestio de aqua de terra	29 766	2.85	2.69
G. Galilei	Siderus Nuncius	73 346	2.88	2.70
D. Alighieri	De Vulgari Eloquentia	82 765	3.02	2.84
G. Galilei	Trattati di fortificazione	121 893	2.58	2.38
E. Salgari	I pirati della Malesia	370 949	2.85	2.45
G. Leopardi	Zibaldone	5 772 133	2.84	2.35

In this way, the number of matches is in general bigger then that of BCL, but the numbers we need to indicate the positions of the matches are usually much smaller (in particular for texts from large alphabets). Let us note that we can minimize some of this numbers: in the example, the last match *bbbaa* can not start with the second *b* from the end of *y* because this *b* is also the last character of a copy of the previous match. So we can specify the position of the new match counting only the possible characters in which the match can start, according to the fact that the previous match is maximal. In this case we can specify the position with the index 3 and not 4.

We codify all this numbers exactly as `gzip` does, and finally, we compress the list of characters and lengths, and the list of the positions, using an arithmetic encoding conditioned to the first character of the match. This leads to the implementation of a true compressor program which effectively shows better compression rates with respect to `gzip`, as shown in Table 11 (see [2]).

The details of this method are designed to optimize the compression ratio in order to obtain a good estimate of the relative entropy between texts. Because of this specific optimization, the complexity of the implementation naturally increases. The interested reader can compare our method with the more simple one introduced by J. Ziv and N. Merhav [21], even if it turns out to be less accurate for our aims: the authors estimate the similarity between the texts x and y with

$$N \, \log |y| \, / \, |x|,$$

where $|y|$ and $|x|$ are the lengths of the two texts, respectively, and N is just the cardinality of the parsing of x in y, as described at the beginning of this section.

This algorithm has been already used for A.A. in [4], but it was also previously tested with the Gramsci corpus used in [2], for which *lzwe* give the same results of BCL.

6.3   The Voting Algorithm

In our specific DP scenario, but also as common to many problems in A.A., we have constructed a reference corpus made of texts with known attribution, assuming/hoping that this corpus contains all stylistic features needed to efficiently discriminate between the authors (in our numerical experiments, either Cima/Montale or the three authors "ma", "mb", "c"). As it is in our case the size of each class can be very different, whereas our methods are both quite sensitive to the length of the reference texts (in particular the entropic one).

In order to construct a reference corpus with an homogenous distribution of lengths of the texts, we split the known texts in pieces of approximately the same length. In this way we have transformed a problem of managing texts of different sizes, in the problem of managing a different number of texts of the same size.

As taken from [4] and to make things a little bit more abstract, just consider a corpus made out of just two authors, A and B with very different character lengths.[7] Given now a size L, we split both A and B in N_A and $N - B$ fragments of basically the same size $\approx L$, respectively. We now analyze an unknown text X by calculating its similarity distance (using either one of the two methods) between the N_A fragments of A and the N_B fragments of B, respectively. Our voting system is just a natural way of turning all this $N_A + N_B$ distances into a scalar index and then into an attribution for X. To be more formal, given X and the ordered set of distances from the fragments of A and B, let us denote the rank of X with $c_1 \ldots c_N$, where $c_i \in \{A, B\}$ is the author of the ith fragment in the rank (from smaller to bigger distance values), and $N = N_A + N_B$. Namely, c_i be the value of the ith position in the ranking, i.e., $c_i = A$ if the ith text is of the author A, and $c_i = B$ elsewhere. Let us consider a text $\tilde{a}$ of the author A. We can model the attribution procedure assuming that similarity distance with a fragment a of author A is a random positive variable $d(\tilde{a}, a)$ in a and that also the distance with a fragment b of the author B is again another random positive variable. With a monotonic transformation of the distances, we can assume that $d(\tilde{a}, a)$ is uniformly distributed in $[0, 1]$. Our main assumption is that with this transformation the other distribution function for the random variable $d(\tilde{a}, b)$ turns into a power law: $P(d(\tilde{a}, b) < z) = z^{1+\beta}$, with $\beta > 0$. Moreover, we also assume that different distance values are independent.

Define

$$\bar{m}_A(k) = \sum_{i=1}^{k} \chi\{c_i = A\} \quad \bar{m}_B(k) = \sum_{i=1}^{k} \chi\{c_i = B\}$$

with $\bar{m}_A(k) + \bar{m}_B(k) = k$.

---

[7]Extension to three authors just involves few more indexes, but it is essentially equivalent.

The value $\bar{m}_A(k)$ is the number of texts of class A which appear in the rank until position k. With these assumptions, we can calculate explicitly the probability of observing the rank $c_1 \ldots c_N$ if $X \in A$. The calculations involves the evaluation of some multiple integrals and yields the formula:

$$P(c_1 c_2 \ldots c_N | X \in A) = (1 + \beta)^{N_B} \frac{1}{\prod_{k=1}^{N} (k + \bar{m}_A(k))}$$

A first order expansion in β yields:

$$P(c_1 c_2 \ldots c_N | X \in A) \simeq \frac{1}{N!} \left(1 + \beta N_B - \beta \sum_{k=1}^{N} \frac{\bar{m}_B(k)}{k} \right).$$

We can now define the I_B and the I_A index as

$$I_B = \sum_{k=1}^{N} \frac{\bar{m}_B(k)}{k} - N_B, \quad \text{and} \quad I_A = \sum_{k=1}^{N} \frac{\bar{m}_A(k)}{k} - N_A$$

namely,

$$P(c_1 c_2 \ldots c_N | X \in A) \simeq \frac{1}{N!} (1 - \beta I_B)$$

We can now make the same assumption for the case $X = \tilde{b} \in B$; in particular assuming the law $z^{1+\gamma}$ for the distribution function of $d(\tilde{b}, a)$ with respect to $d(\tilde{b}, b)$, we obtain:

$$P(c_1 c_2 \ldots c_N | X \in B) \simeq \frac{1}{N!} (1 - \gamma I_A)$$

Note that $I_A + I_B = 0$. Now we can use a maximum likelihood method (which in this case corresponds to a bayesan method with the dubious texts equidistributed between B and A) and we can attribute X to B if

$$P(c_1 c_2 \ldots c_N | X \in B) > P(c_1 c_2 \ldots c_N | X \in A) \Leftrightarrow -\gamma I_A > -\beta I_B \Leftrightarrow I_B > 0$$

In the same way, we choose attribution A if $I_A = -I_B > 0$.

References

1. Altmann, E.G., Cristadoro, G., Degli Esposti, M.: On the origin of long-range correlations in texts. Proc. Natl. Acad. Sci. **109**, 11582–11587 (2012)
2. Basile, C., Benedetto, D., Caglioti, E., Degli Esposti, M.: An example of mathematical authorship attribution. J. Math. Phys. **49**, 1–20 (2008)

3. Benedetto, D., Caglioti, E., Loreto, V.: Language trees and zipping. Phys. Rev. Lett. **88**(4), 48702 (2002)
4. Benedetto, D., Degli Esposti, M., Maspero, C.: The puzzle of Basils Epistula 38: a mathematical approach to a philological problem. J. Quant. Linguist. **20**(4), 267–287 (2013)
5. Bennet, W.R.: Scientific and Engineering Problem-Solving with the Computer. Prentice-Hall, Englewood Cliffs (1976)
6. Canettieri, P., Italia, P.: Un caso di attribuzionismo novecentesco: il Diario Postumo di Montale. Cogn. Philol. **6** (2013)
7. Clement, R., Sharp, D.: Ngram and Bayesian classification of documents. Lit. Linguist. Comput. **18**, 423–447 (2003)
8. Condello, F.: I filologi e gli angeli. E' di Eugenio Montale il Diario postumo?. Bononia University Press, ISBN-13: 978-8873959786 (2014)
9. Cover, T.M., Thomas, J.A.: Elements of Information Theory. Wiley-Interscience, Hoboken (2006)
10. Juola, P.: Authorship attribution. FNT Inf. Retr. **1**, 233–334 (2007)
11. Kešelj, V., Peng, F., Cercone, N., Thomas, C.: N-gram-based author profiles for authorship attribution. In: Kešelj, V., Endo, T. (eds.) Proceedings of the Conference Pacific Association for Computational Linguistics, PACLING'03, pp. 255–264. Dalhousie University, Halifax (2003)
12. Lempel, A., Ziv, J.: On the complexity of finite sequences. IEEE Trans. Inf. Theory **IT–22**(1), 75–81 (1976)
13. Lempel, A., Ziv, J.: A universal algorithm for sequential data compression. IEEE Trans. Inf. Theory **23**(3), 337–343 (1977)
14. Lempel, A., Ziv, J.: Compression of individual sequences via variable-rate coding. IEEE Trans. Inf. Theory **IT–24**(5), 530–536 (1978)
15. Mendenhall, T.C.: The characteristic curves of composition. Science **9**(214), 237–249 (1887)
16. Proceedings of the Workshop, A carte scoperte. Eugenio Montale. E' il "Diario Postumo" un falso ?, Bologna, 11 novembre 2014. Bononia University Press (to appear, 2016)
17. Stamatatos, E.: A survey of modern authorship attribution methods. J. Am. Soc. Inf. Sci. Technol. **60**, 538–556 (2009)
18. Stamatatos, E., Daelemans, W., Verhoeven, B., Potthast, M., Stein, B., Juola, P., Sanchez-perez, M.A., Barrón-cedeño, A.: Overview of the author identification task at PAN-2013. Notebook Papers of CLEF 2013 LABs and Workshops (CLEF-2013) (2013)
19. Shannon, C.E.: A mathematical theory of communication. Bell Syst. Tech. J. **27**, 379–423, 623–656 (1948)
20. Wyner, A.D., Ziv, J., Wyner, A.J.: On the role of pattern matching in information theory. IEEE Trans. Inf. Theory **44**(6), 2045–2056 (1998)
21. Ziv, J., Merhav, N.: A measure of relative entropy between individual sequences with application to universal classification. IEEE Trans. Inf. Theory **39**(4), 1270–1279 (1993)

Universality and Creativity: The Usage of Language in Gender and Irony

Paolo Rosso, Delia Irazú Hernández Farías and Francisco Rangel

Abstract Author profiling deals with distinguishing between classes of authors rather than individual authors on the basis of their usage of language. What is much more subjective in terms of usage of language is when authors employ irony as linguistic device. The aim of this paper is to introduce the reader to concepts such as universality of language among classes of authors, e.g. of the same gender, and creativity in irony.

1 Universality of Language in Gender

The aim of author profiling is to identify the linguistic profile of an author on the basis of the writing style. Author profiling tries to determine the author's gender, age, native language, personality type, etc. by analysing her texts. The writing style reflects the profile of the author, who decides, often unconsciously, how to choose and combine words. Profiling anonymous authors is a problem of growing importance, both from forensic and marketing perspectives. The study of how certain linguistic features vary according to the profile of the authors is a subject of interest for several different areas such as psychology, linguistics and more recently, computational linguistics. In the following, we briefly describe some state-of-the-art approaches, including those of the research teams that participated in the author profiling shared

P. Rosso (✉) · D.I. Hernández Farías · F. Rangel
Natural Language Engineering Lab, Universitat Politècnica de València,
Valencia, Spain
e-mail: prosso@dsic.upv.es

D.I. Hernández Farías
e-mail: dhernandez1@dsic.upv.es

F. Rangel
Autoritas Consulting S.A., Valencia, Spain
e-mail: francisco.rangel@autoritas.es

© Springer International Publishing Switzerland 2016   177
M. Degli Esposti et al. (eds.), *Creativity and Universality in Language*,
Lecture Notes in Morphogenesis, DOI 10.1007/978-3-319-24403-7_11

task that has been organised in 2013 and 2014 at the PAN Lab at CLEF,[1] focussing especially on gender identification.

Author profiling has become a hot research topic and many are the research works that have been carried out recently. In this section, we briefly describe some of the most well-known ones in order to introduce the reader to author profiling, and more concretely, to gender identification. Argamon et al. [1] analysed formal written texts extracted from the British National Corpus, combining function words with part-of-speech features. Koppel et al. [17] studied the problem of automatically determining an author's gender in social media by proposing combinations of simple lexical and syntactic features. Schler et al. [35] studied the effect of gender, and age, in the writing style in blogs; they gathered over 71,000 blogs and obtained a set of stylistic features like non-dictionary words, parts-of-speech, function words and hyperlinks, combined with content features, such as word unigrams with the highest information gain. In the three previous research works, the authors achieved an accuracy of approximately 80 % in gender prediction. Goswami et al. [14] added some new features as slang words and the average length of sentences, improving accuracy to 89.2 % in gender detection. More recently, Nguyen et al. [21] measured the effect of the gender in the performance of age identification, considering both variables as inter-dependent, and achieved correlations up to 0.74 and mean absolute errors between 4.1 and 6.8 years. Pennebaker et al. [25] connected language use with personality traits, studying how the variation of linguistic characteristics in a text can provide information regarding the gender and age of its author.

In the author profiling task at PAN 2013 [29], participants approached the task of identifying gender and age in a large corpus collected from social media. At PAN 2014 [30], four corpora of different genres were considered, both in English and Spanish: social media, blogs, Twitter and hotel reviews. Systems using simple content features, such as bag-of-words or word n-grams achieved very good accuracies. Readability measures (e.g. Gunning Fog index, Flesch–Kinkaid, etc.) were also used. The best results were obtained using a second-order representation based on relationships between documents and profiles [23]. With respect to gender identification, the highest accuracy was obtained in Twitter for English (73 %) and in social media texts for Spanish (68 %).

In a couple of recent works [28, 31], we investigated the impact of emotions on author profiling. Concretely we investigated style-based and emotion-labelled graph (EmoGraph) features. On the basis of what was already investigated for English by the authors such as Pennebaker [25], we aimed to investigate the use of the different morphosyntactic categories in Spanish. In our first style-based approach (Rangel-S) we obtained frequencies, punctuation marks and emoticons using regular expressions, whereas the morphosyntactic categories were obtained with the Freeling library.[2] Moreover, a Spanish Emotion Lexicon (SEL) [10] was employed to consider also the usage of emotions. In our second EmoGraph approach,

---

[1]PAN Lab on Uncovering Plagiarism, Authorship, and Social Software Misuse: http://pan.webis.de.

[2]http://nlp.lsi.upc.edu/freeling/.

Table 1 Results in accuracy for gender identification in PAN-AP13 corpus (Spanish)

Ranking	Team	Accuracy
1	Santosh	0.6473
2	**EmoGraph**	0.6365
3	Pastor	0.6299
4	Haro	0.6165
5	Ladra	0.6138
...	...	
8	**Rangel-S**	0.5713
...	...	
18	Baseline	0.5000
...	...	
23	Gillam	0.4784

we modelled each part-of-speech as a node of the graph, and each edge defines the sequence of parts-of-speech in the text. The obtained graph has been enriched with semantic and affective information. For each word, topics were obtained with the help of Wordnet Domains.[3] Adjectives, verbs and adverbs were annotated with their polarity [16] and associated emotions [10]. Verbs were annotated within the following categories [18]: (i) perception (see, listen, smell...); (ii) understanding (know, understand, think...); (iii) doubt (doubt, ignore...); (iv) language (tell, say, declare, speak...); (v) emotion (feel, want, love...); (vi) and will (must, forbid, allow). We carried out some experiments with the Spanish partition of the PAN-AP-13 corpus. The best results obtained employing support vector machine are shown in Table 1. EmoGraph outperforms significantly better than Rangel-S in gender identification. The t-student test showed no significant difference (at 95 % of confidence) between EmoGraph and the method ranked first. Therefore, EmoGraph shows the feasibility of approaching gender identification considering emotions and their usage in text.

As mentioned before, in EmoGraph the obtained graph has been enriched with affective (SEL) and semantic information. To this aim the semantic classification of verbs was considered. On the basis of what was investigated in [18], 158 verbs have been manually annotated with one of the following semantic categories: perception (see, listen, smell...); understanding (know, understand, think...); doubt (doubt, ignore...); language (tell, say, declare, speak...); emotion (feel, want, love...); and will (must, forbid, allow...). In terms of usage of verbs, Fig. 1 illustrates that females use more *emotional* verbs (e.g. feel, want, love...) than males, who use more *language* verbs (e.g. tell, say, speak...) [31].

In order to investigate whether the usage of language changes or not in females versus males; in [32] we analysed the distribution grammatical categories in a dataset of 1,200 Facebook comments in Spanish. Statistics are shown in Table 2. We can appreciate some important variations in the usage of the grammatical categories by

---

[3]http://wndomains.fbk.eu.

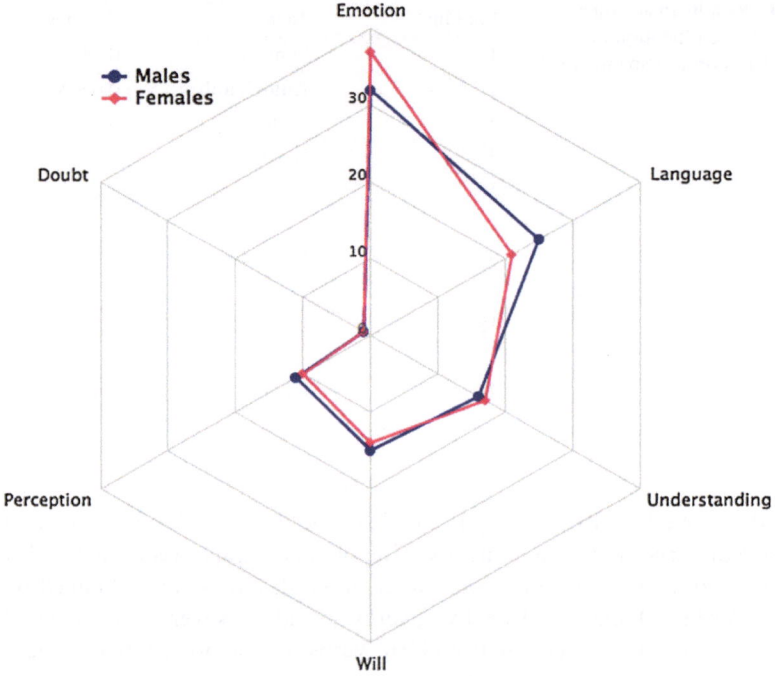

Fig. 1 Use of verb types per gender in PAN-AP13 corpus

Table 2 Distribution of grammatical categories by gender

POS	ALL	Female	Male
ADJ	6.49	6.45	6.53
ADV	3.93	3.91	3.94
CONJ	9.51	9.46	9.55
DET	7.25	**7.74**	6.81
INTJ	0.23	**0.30**	0.18
PREP	6.06	5.85	**6.25**
PRON	2.45	**2.67**	2.24
NOM	31.89	31.53	32.21
VERB	15.38	15.32	15.44

gender. For instance, as found for English [26], we also verified for Spanish that men use more prepositions than women (+6.84 %) and women use more pronouns (+19.20 %), determinants (+13.66 %) and interjections (+66.67 %) than men.

2 Creativity of Language in Irony

In [27] we investigated the usage of irony in a Facebook Spanish dataset.[4] We calculated the Fleiss' Kappa (kFleiss) to measure the inter-annotator agreement with respect to irony annotation. kFleiss is a statistical measure that allows to consider more than two annotators (we have three annotators) [11]. A kFleiss=1 is obtained when the raters are in complete agreement, otherwise if there is no agreement kFleiss=0. We obtained a kFleiss value equal to 0.0989; i.e. a very low index of agreement. This shows how creative and subjective irony is perceived.

Some examples of language creativity in ironic comments are shown below. Comments are shown in their original language in order to preserve their ironic sense. We provide an English explanation based on our own interpretation, in order to show the difficulty of the task.

e.g. *"Pitbul es cultura, £no ves que te enseña a contar? aunque sea sólo hasta 3"*

In the previous comment, the author criticises the singer for including in his lyrics "one, two, three...". The author says that this is culture because listening such singer, anyone can count. At least, until number three. The author expresses a positive comment using a remark in order to emphasising his negative opinion about this singer. In this comment, two out of three annotators agreed on annotating this comment as ironic.

e.g. *"Que viva, pero muy lejos!"*

In this comment, the author expresses her intention of being far from someone, mentioning at first a positive desire and, finally, showing her real intention. All the annotators perceived the comment as ironic.

e.g. *"Pobres, en el fondo producis ternura...que triste tiene que ser haber votado al PP."*

In this comment, the author expresses shame towards people. The author's remark is about people's judgement for voting for the right wing Spanish party. The author expresses a negative comment in order to show her real intention. Two of the three annotators tagged this comment as ironic.

e.g. *"Eres muy injusto y quiero que sepas que la infanta cuando se fue a vivir a su nueva vivienda recien reformada y a pesar de ser mucho mas pequeña que la zarzuela se mudo convencida de que era una VPO o no?..."*

In the previous comment, the author says that the former Spanish King's daughter moved to a new residence. She says that the Spanish King's daughter was convinced that this new house is a kind of state subsidy housing because it is smaller than Zarzuela's Palace, the Residence of the Spanish royal family. The author expresses a positive remark about someone's judgement including comparisons in order to emphasise the utterance's ironic sense. The annotators tagged this comment as ironic.

e.g. *"Yo soy presunta ciudadana española y digo esto porque no estoy segura de si realmente lo soy o si vivo en una realidad paralela donde nuestro presi es más inútil que una neurona de Paris Hilton."*

---

[4]EmIroGeFB: http://ow.ly/uQWEs.

In the last comment, the author alludes the possibility of living in a parallel reality because her country is governed for someone useless than a Paris Hilton's neuron. The author compares two remarks in the same comment, in order to emphasise her real intention to show disagreement with the government of her country. All the annotators perceived the comment as ironic.

Above we showed how far from being universal irony is, being its usage very much subjective and depending on people, their linguistic and cultural contexts, their moods etc. This makes irony difficult to be perceived as, therefore, difficult to be detected, even more if automatically. Automatic irony detection has attracted the attention of researchers from both machine learning and natural language processing [34, 39]. A shared task on sentiment analysis of figurative language in Twitter has been organised at SemEval 2015[5] [13] and for Italian at Evalita 2014 [3].

Reyes et al. [33] address the problem of irony detection as a classification task; the authors proposed a model organized according to four types of conceptual features: signatures, unexpectedness, style and emotional scenarios. Buschmeier et al. [8] present an analysis of 29 features (like punctuation marks and emoticons) previously applied in irony detection research; their main goal is to investigate the impact that has the features elimination on the performance of their approach. A survey that includes both philosophical and literary works investigating ironic communication and some computational efforts to operationalize irony detection is presented by Wallace in [39]. Barbieri and Saggion [4] used six groups of features (frequency, written-spoken, intensity, structure, sentiments, synonyms, ambiguity) in order to classify ironic tweets; they used the same dataset that [33]. Veale and Hao [38] presented a linguistic approach to separate irony from non-irony in figurative comparisons over a corpus of web-harvested similes. Bosco et. al. in [6] presented a study that investigates sentiment and irony in online political discussion social media in Italian.

In a recent work [15], we addressed irony detection as a classification problem, considering both surface patterns (such as punctuation marks, emoticons) and novel sentiment analysis features (Sentiment Score and Polarity Value, using Hu and Liu's [16] and AFINN [20] lexicons, respectively) in order to take advantage of resources that allow to measure the overall sentiment expressed in each tweet. Table 3 shows the obtained improvement in terms of F-measure with respect of state-of-the-art methods on the same Twitter corpus.

Another case of where creativity in language is employed is sarcasm. For both computational and non-computational linguistic purposes, most of the time irony and sarcasm are often viewed as the same figurative language device. The question of whether irony and sarcasm are separated or similar linguistic phenomena is a controversial issue in the literature and no clear consensus has already been reached [12]. Irony is often considered as an umbrella term that covers also sarcasm.

We are interested in investigating the differences between sarcasm and irony in social media. An analysis of similarities and differences between ironic and

---

[5]Given a set of tweets the task consist in determining whether the user has expressed a positive, negative or neutral sentiment; more information is available at: http://alt.qcri.org/semeval2015/task11/.

Table 3 Results of our model against the ones obtained by Reyes et al. and Barbieri and Saggion

	Irony versus		
	Education	Humor	Politics
Reyes et al.	0.70	0.76	0.73
Barbieri and Saggion	0.73	0.75	0.75
Hernández-Farías et al.	**0.78**	**0.79**	**0.79**

sarcastic tweets has been presented by Wang [40]. In [37] we carried out a distribution and correlation analysis in order to detect the main characteristics with respect to structural, psycholinguistic and emotional features over a set of tweets rich in figurative language content. We used the corpora of the Task 11 of Semeval-2015.[6] In our analysis, we incorporated information about the psychological and emotional content of tweets by means of not only sentiment lexicons (AFINN [20], Hu and Liu's lexicon [16], General Inquirer [36], SentiWordNet [2] and SenticNet [9]), but also a variety of psycholinguistic resources enabling the analysis of affective content from different perspectives (ANEW [7], DAL [41], LIWC [24] and EmoLex [19]). We considered a score of each resource that is calculated as the frequency of matches between words that compose each tweet and the corresponding lexicon.

On the basis of the previous analysis we selected a set of features (structural, psycholinguistic sentiment scores and emotions) in order to see if they can help to distinguish irony from sarcasm. According to our results, sarcastic messages contain more positive words than ironic ones (taking into account the positive and negative words in General Inquirer, Hu and Liu's lexicon and SentiWordNet). Furthermore, we observe evident traces in the values of ANEW and DAL affective lexicons. Higher values for irony confirm that this device is more subtle and less direct. On the other side, lower values for sarcasm confirm that is more direct and less creative than irony.

The irony-sarcasm identification task has been formulated in terms of a binary classification task #irony versus #sarcasm. We evaluated our approach over a dataset composed by 20,000 tweets, that were compiled using a subset of 10,000 tweets with the hashtag #irony and 10,000 tweets containing #sarcasm hashtag retrieved by [5]. We applied a set of classification algorithms composed by: Naïve Bayes, decision tree and support vector machine.[7] We perform a tenfold cross-validation and report F-measure in Table 4. The results obtained are better than those reported in the state-of-the-art on the same dataset.

---

[6]http://alt.qcri.org/semeval2015/task11.

[7]We use Weka toolkit's version of each classifier available at: http://www.cs.waikato.ac.nz/ml/weka/dowloading.html.

Table 4 #irony versus #sarcasm

F-measure	#irony versus #sarcasm
Naïve Bayes	0.646
Decision tree	0.642
Support vector machine	**0.677**

3 Conclusions

In this paper we have approached universality in language from the perspective of the usage of language among people with the same gender profile, i.e. belonging to the same class (female versus male). As example of creativity in language, the case of irony was described. Imaginative language is also commonly used in deception, for instance in deceptive opinions, where more verbs, adverbs and pronouns are used [22]. The relation between deceptive and imaginative language is an intriguing issue to be investigated further.

Acknowledgments The research work was carried out in the VLC/CAMPUS Microcluster on Multimodal Interaction in Intelligent Systems in the framework of the SomEMBED TIN2015-71147-C2-1-P MINECO research project and under the Generalitat Valenciana grant ALMAMATER (PrometeoII/2014/030). The National Council for Science and Technology (CONACyT-Mexico) has funded the research work of the second author (Grant No. 218109/313683, CVU-369616). The work of the third author was partially funded by Autoritas Consulting SA and by Ministerio de Economia de España under grant ECOPORTUNITY IPT-2012-1220-n430000.

References

1. Argamon, S., Koppel, M., Fine, J., Shimoni, A.R.: Gender, genre, and writing style in formal written texts. In: TEXT, vol. 23, pp. 321–346 (2003)
2. Baccianella, S., Esuli, A., Sebastiani F.: SentiWordNet 3.0: an enhanced lexical resource for sentiment analysis and opinion mining. In: Proceedings of the Seventh International Conference on Language Resources and Evaluation (LREC) (2010)
3. Basile, V., Bolioli, A., Nissim, M., Patti, V., Rosso, P.: Overview of the Evalita 2014 SENTIment POLarity Classification Task. In: Proceeding of the 4th Evaluation Campaign of Natural Language Processing and Speech tools for Italian, EVALITA-2014, Pisa, Italy, pp. 50-57, Dec. 9-11 (2014)
4. Barbieri, F., Saggion, H.: Modelling irony in twitter. In: Proceedings of the Student Research Workshop at the 14th Conference of the European Chapter of the Association for Computational Linguistics, pp. 56-64 Association for Computational Linguistics (2014)
5. Barbieri, F., Saggion, H., Ronzano, F.: Modelling sarcasm in twitter, a novel approach. In: Proceedings of the 5th Workshop on Computational Approaches to Subjectivity, Sentiment and Social Media Analysis, pp. 50–58. Association for Computational Linguistics (2014)
6. Bosco, C., Patti, V., Bolioli, A.: Developing corpora for sentiment analysis: the case of irony and Senti-TUT. IEEE Intell. Syst. **28**(2), 55–63 (2013)
7. Bradley, M., Lang, P.: Affective Norms for English Words (ANEW): Instruction Manual and Affective Ratings (1999)

8. Buschmeier, K., Cimiano, P., Klinger, R.: An impact analysis of features in a classification approach to irony detection in product reviews. In: Proceedings of the 5th Workshop on Computational Approaches to Subjectivity, Sentiment and Social Media Analysis, WASSA-2014, pp. 42-49. Association for Computational Linguistics (2014)

9. Cambria, E., Havasi, C., Hussain, A.: SenticNet 2: a semantic and affective resource for opinion mining and sentiment analysis. In: Proceedings of the FLAIRS: Florida Artificial Intelligence Research Society Conference (2012)

10. Díaz Rangel, I.: Detección de afectividad en texto en español basada en el contexto lingüístico para síntesis de voz. Tesis Doctoral. Instituto Politécnico Nacional. México (2013) (in Spanish)

11. Fleiss, Joseph L.: Measuring nominal scale agreement among many raters. Psychol. Bull. **76**(5), 378–382 (1971)

12. Gibbs, R.W., Colston, H.L.: Irony in Language and Thought. Routledge (Taylor and Francis), New York (2007)

13. Ghosh, A., Li, G., Veale, T., Rosso, P., Shutova, E., Reyes, A., Barnden, J.: SemEval-2015 Task 11: sentiment analysis of figurative language in twitter. In: Proceedings International Workshop on Semantic Evaluation (SemEval-2015), Co-located with NAACL and *SEM (2015)

14. Goswami, S., Sarkar, S., Rustagi, M.: Stylometric analysis of bloggers' age and gender. In: Proceedings of the Third International Conference on Weblogs and Social Media (ICWSM). AAAI Press (2009)

15. Hernández-Farías, I., Benedí, J.M., Rosso, P.: Applying basic features from sentiment analysis for automatic irony detection. In: Iberian Conference on Pattern Recognition and Image Analysis (IbPRIA) 2015, Santiago de Compostela (Spain), June 17-19 (2015)

16. Hu, M., Liu, B.: Mining and summarizing customer reviews. In: Proceedings of the Tenth ACM SIGKDD International Conference on Knowledge Discovery and Data Mining (2004)

17. Koppel, M., Argamon, S., Shimoni, A.: Automatically categorizing written texts by author gender. Lit. Linguist. Comput. **17**(4), 401–412 (2003)

18. Levin, B.: English Verb Classes and Alternations. University of Chicago Press, Chicago (1993)

19. Mohammad, S.M., Turney, P.D.: Crowdsourcing a word-emotion association lexicon. Comput. Intell. **29**(3), 436–465 (2013). Wiley Online Library

20. Nielsen, F.: A new ANEW: Evaluation of a word list for sentiment analysis in microblogs. In: Proceedings of the Workshop on Making Sense of Microposts (2011)

21. Nguyen, D., Gravel, R., Trieschnigg, D., Meder, T.: How Old Do You Think I Am?; a study of language and age in twitter. In: Proceedings of the Seventh International AAAI Conference on Weblogs and Social Media (2013)

22. Ott, M., Choi, Y., Cardie, C., Hancock, J. T.: Finding deceptive opinion spam by any stretch of the imagination. In: Proceedings of the 49th Annual Meeting of the Association for Computational Linguistics: Human Language Technologies, Portland, Oregon, USA (2011)

23. Pastor Lopez-Monroy, A., Montes-Gomez, M., Jair Escalante, H., Villasenor-Pineda, L., Villatoro-Tello, E.: INAOEs participation at PAN13: author profiling task. In: Notebook for PAN at CLEF (2013)

24. Pennebaker, J.W., Francis, M., Booth, R.: Linguistic inquiry and word count: LIWC 2001. In: Mahway, vol. 71. Lawrence Erlbaum Associates (2001)

25. Pennebaker, J.W., Mehl, M.R., Niederhoffer, K.: Psychological aspects of natural language use: our words, our selves. Ann. Rev. Psychol. **54**, 547–577 (2003)

26. Pennebaker, J.W.: The Secret Life of Pronouns: What Our Words Say About Us. Bloomsbury Press, London (2011)

27. Rangel, F., Hernández, I., Rosso, P., Reyes, A.: Emotions and irony per gender in facebook. In: Proceedings of Workshop on Emotion, Social Signals, Sentiment & Linked Open Data (ES3LOD), LREC-2014, Reykjavík, Iceland, May 26–31 (2014)

28. Rangel, F., Rosso, P.: On the identification of emotions and authors' gender in facebook comments on the basis of their writing style. In: Proceedings of ESSEM Workshop on Emotion and Sentiment in Social and Expressive Media, AIxIA, vol. 1096, pp. 34-46. http://CEUR-WS.org (2013)

29. Rangel, F., Rosso, P., Koppel, M., Stamatatos, E., Inches, G.: Overview of the author profiling task at PAN 2013. In: Forner P., Navigli R., Tufis D. (eds.), CLEF 2013 Labs and Workshops, Notebook Papers, vol. 1179, Valencia, Spain, Sept. 23–26. http://CEUR-WS.org (2013)
30. Rangel, F., Rosso, P., Chugur, I., Potthast, M., Trenkmann, M., Stein, B., Verhoeven, B., Daelemans, W.: Overview of the 2nd author profiling task at PAN 2014. In: Cappellato, L., Ferro, N., Halvey, M., Kraaij, W. (eds.) CLEF 2014 Labs and Workshops, Notebook Papers, vol. 1180, pp. 898-827. http://CEUR-WS.org (2014)
31. Rangel, F., Rosso, P.: On the impact of emotions on author profiling. In: Information Processing & Management **52**(1), 73–92 (2016)
32. Rangel, F., Rosso, P.: Use of language and author profiling: identification of gender and age. In: 10th International Workshop on Natural Language Processing and Cognitive Sciences NLPCS 2013 CIRM, Marseille, France, Oct. 13–17 (2013)
33. Reyes, A., Rosso, P., Veale, T.: A multidimensional approach for detecting irony in twitter. Lang. Res. Eval. **47**(1), 239–268 (2013)
34. Reyes, A., Rosso, P.: On the difficulty of automatically detecting irony: beyond a simple case of negation. Knowl. Inf. Syst. **40**(3), 595–614 (2014)
35. Schler, J., Koppel, M., Argamon, S, Pennebaker, J.W.: Effects of age and gender on blogging. In: AAAI Spring Symposium: Computational Approaches to Analyzing Weblogs, pp. 199-205. AAAI (2006)
36. Stone, P.J., Hunt, E.B.: A computer approach to content analysis: studies using the general Inquirer system. In: Proceedings of the May 21–23, Spring Joint Computer Conference (1963)
37. Sulis, E., Hernández-Farias, I., Rosso, P., Patti, V., Ruffo, G.: Figurative messages and affect in twitter: differences between #irony, #sarcasm and #not, Knowledge-Based Systems (submitted)
38. Veale, T., Hao, Y.: Detecting Ironic Intent in Creative Comparisons. In: Coelho, H., Studer, R.,Wooldridge, M. (eds.), Proceedings of the 19th European Conference on Artificial Intelligence (ECAI) Frontiers in Artificial Intelligence and Applications, vol. 215, pp. 765-770. IOS Press (2010)
39. Wallace, B.: Computational irony: a survey and new perspectives. In: Artificial Intelligence Review, pp. 1-17. Springer Netherlands (2013)
40. Wang, A.P.: #Irony or #Sarcasm—a quantitative and qualitative study based on twitter. In: Proceedings of the PACLIC: the 27th Pacific Asia Conference on Language, Information, and Computation (2013)
41. Whissell, C.: Using the revised dictionary of affect in language to quantify the emotional undertones of samples of natural languages. In: Psychological Reports (2009)

Computational Approaches to the Analysis of Human Creativity

Fabio Celli

Abstract In this paper we address the issue of creativity and style computation from a natural language processing perspective. We introduce a computational framework for creativity analysis with two approaches, one agnostic, based on clustering, and one knowlegde-based, that exploits supervised learning and feature selection. While the agnostic approach can reveal the uniqueness of authors in a meaningful context, the knowledge-based approach can be exploited to extract the culturally relevant features of works and to predict social acceptance. In both the approaches, it is required a great effort to define symbols to represent meaningful cues in creativity and style.

1 Introduction and Related Work

Understanding human creativity is a long-standing issue that poses great challenges to many disciplines, humanities above all, but also psychology and computer science. On the humanities side, the flow theory [1, 2] defined creativity as the process by which a symbolic domain in the culture is changed. This theory emphasizes three elements:

(1) a culture that defines rules;
(2) an entity, person or event, who brings novelty into the rules;
(3) a set of people that recognize, accept and validate the innovation.

From the psychological side, the attempts to develop a creativity quotient test similar to the intelligence quotient (IQ) have been unsuccessful [3]. Guilford [4] and Torrance [5] developed tests of creative thinking from the 1960s to the 1980s, scoring the uniqueness of subjects in different tasks. These tests have been strongly criticized in recent years [6], with the argument that a person's creativity can only be assessed indirectly, for example with observer ratings [7]. This method has the great problem of the agreement between raters and the limited context that can be analyzed. One

F. Celli (✉)
University of Trento, Trento, Italy
e-mail: fabio.celli@unitn.it

© Springer International Publishing Switzerland 2016   187
M. Degli Esposti et al. (eds.), *Creativity and Universality in Language*,
Lecture Notes in Morphogenesis, DOI 10.1007/978-3-319-24403-7_12

of the self assessments available is the creativity achievement questionaire [8], that is based on the idea that a person's creativity can be measured by the person's achievements recognized by the society in different fields of creativity, from art and music to business. In computer science, most of the effort in this field has been done by artificial creativity, that focuses on social aspects. Based on the dual generate-and-test model of creativity [9], that sees the creative process as the interaction of individuals with their social environment, artificial creativity promotes the study of the creative behaviour of individuals in societies by means of networks of agents, whose parameters and interactions can be observed in a controlled setting [10]. Other computational approaches that focus more on the creative process include logic and machine learning. Wiggins [11] proposed a finite-state language to formalize creativity as a set of functions that generate new elements from the existing ones, while Barbieri [12] successfully exploited a constrained markov process to generate lyrics in the style of an existing author.

We summarize the different approaches and views of creativity in four types:

(1) **Nomothetic versus idiographic view**. The nomothetic approach focuses on universals of creativity while the idiographic focuses on individual differences and uniqueness of creative individuals [4, 13].

(2) **Mental versus social phenomenon**. Creativity is seen as a mental or psychological phenomenon for the generation of new ideas or as the result of an audience's appreciation of a novel idea [10].

(3) **Improbabilist versus impossibilist approach**. The improbabilist approaches see creativity as a novel combination of familiar ideas while the impossibilists consider creativity as new ideas never appeared before. This is possible thanks to a transformation of the conceptual space [14].

(4) **Rational versus irrational process**. The rational approaches include logic [15] and evolution. Logic sees creativity in processes like abduction (guessing a conclusion given some premises), eduction (relate properties of different objects), deduction and induction (relate properties from general to specific and viceversa). Evolutionary approaches instead see creativity as an adaptation to the environment motivated by problem solving [16, 17] while the irrational approach focuses on the emotional part of creativity [18].

In this paper, we address the issue of computational Human Creativity Analysis (HCA) from a Natural Language Processing (NLP) perspective. We define the framework for exploiting NLP techniques for HCA, focusing on styles and their relationships with creativity. In particular, we identify two methods to tackle the issue of HCA from two different sides: the first one is social, idiographic, rational and improbabilist (we will call this one "agnostic"); the second one is mental, nomothetic and irrational (we will call the approach to this point of view "knowledge-based"). Based on two different NLP techniques, we develop a framework for HCA, describing the challenges and potential problems, both from a theorethical and practical point of view. The main contributions of this paper to the research community are: (1) the definition of a framework for computational HCA from a quite new perspective and (2) the application of NLP techniques suitable to address HCA tasks.

The paper is structured as follows: in Sect. 2 we define the notions of style and creativity, and we introduce the framework for HCA. In Sect. 3 we will run some experiments with the discussion of the results and in Sect. 4 we will draw some conclusions.

2   The HCA Framework

Our framework for HCA has two different approaches: one agnostic, that measures creativity as uniqueness and one knowledge-based, that measures social acceptance. The agnostic approach adopts unsupervised learning [19], in particular simple K-means algorithms [20]; while the knowledge-based approach exploits supervised learning [21], feature selection methods [22, 23] and annotation, that can be manual or automated by means of knowledge bases.

The basic assumption of the agnostic approach is that styles and creativity can be defined by the contexts where they appear. In order to turn HCA into a learning problem, we define **style** as a set of features that can be extracted from documents of an author, and represented as sequences of symbols, such as bare words, rhyme schemas, notes etc. Creativity is strongly related to style: we define **creativity (uniqueness)** as the distance between a target author and its context in a style feature space, as in Fig. 1.

In this way, creativity can be computed with simple k-means clustering, that is generally a very fast technique and do not require annotation, but just a post-hoc evaluation. Turning to Csikszentmihaly's theoretical framework mentioned in Sect. 1, what we call context is a representation of the culture that defines rules. In this agnostic approach, creativity is seen as a measure of the novelty or uniqueness of the style of an author in a context. The agnostic approach to HCA addresses points 1 and 2 of the Flow theory:

(1) a culture that defines rules, represented as the centroid of the cluster;
(2) a person or event who brings novelty into the rules, represented as the points in the space (authors). The distance between each point and the centroid (accepted culture) is the measure of the novelty of that author with respect to the contextual culture.

In the HCA framework, the third point of the Flow theory—the issue of the set of people that recognize, accept and validate the innovation—can be tackled with the knowledge-based approach by means of feature selection and supervised learning. If we link an appreciation score, retrieved from knowledge bases, to an author's style in a context, we can use it as the target class and predict appreciation from the space

Fig. 1 Representation of the agnostic approach to HCA. Style/feature extraction and comparison between authors are turned into a clustering task. Style is extracted as a set of features from many documents of the same author, then a feature space is used for the inter-author comparison in a clustering task. We define the distance between a target author and the cluster centroid as a measure of creativity

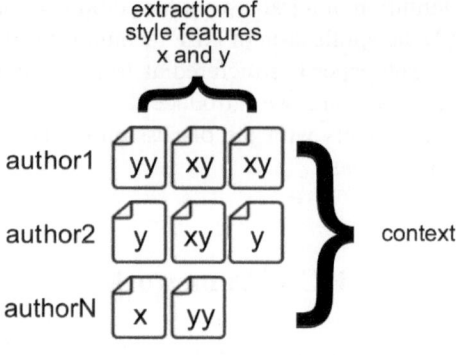

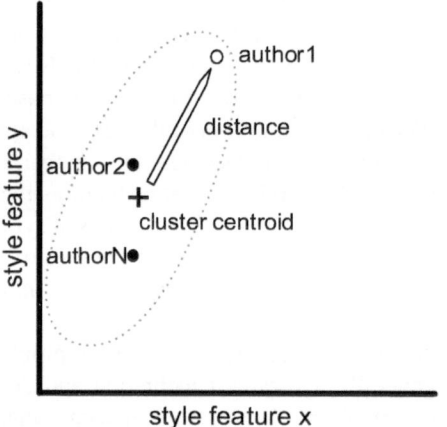

of style features by means of supervised algorithms. These algorithms can provide acceptance models, that link style features to the appreciation scores. In this approach **acceptance** is a function that maps a style feature vector to appreciation scores, that can be extracted from knowledge bases, see Fig. 2.

The heuristic power of our HCA framework depends on the control of two factors:

(1) the definition of a meaningful context;
(2) the definition of meaningful symbols to represent styles.

For this reason, corpus collection is very important. The context collected must be homogeneous, in order to represent the rules defined by the culture as the cluster centroid. For example one can compare authors in the same timespan: in that case the cluster centroid represents the general style, or "taste" of the time. Another important aspect is data representation: we can have as input just simple text or more abstract

Fig. 2 Representation of knowledge-based approach to HCA. Style features are extracted from documents of one or more author(s), and linked to appreciation scores extracted from knowledge bases. Acceptance models, retrieved by means of feature selection methods and supervised machine learning techniques, can be used to predict appreciation scores

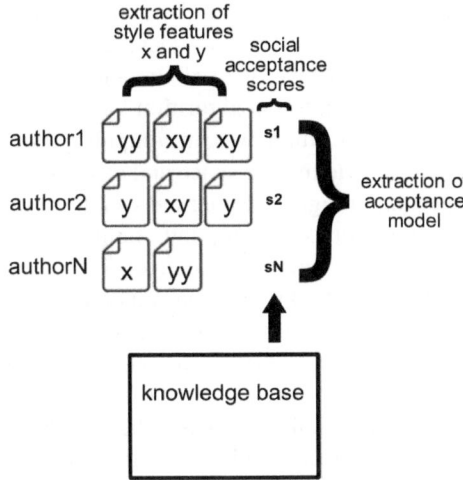

levels of representations, that can be annotated by hand or with ad-hoc algorithms. Language is one clear example of level representation: text can be represented at the level of bare words, part-of-speech, syntactic chunks etc. In the case of style and creativity extraction, formalization is an open issue. For example it is possible to formalize poetry either with rhyme schemas and bare words, or songs as sequences of notes, words, beats-per-minutes, arrangement structures and emotions [24]. In the next section, we will present two sample experiments, one of the agnostic approach and one of the knowledge based.

3 Experiments

We ran two experiments in Weka [25] in order to provide examples of creativity analysis. We collected a corpus of titles of albums and songs of 24 bands formed between 1950s ans 1970s (our context). The corpus contains text and dates for a total of about 19500 tokens. We used the occurrence of words associated to each band as features. We removed stop-words, like prepositions and articles, and built a style feature space of more than 1000 dimensions, where each dimension is the frequency of a word. We limited the words used as features to 1000. We selected to sample the data of only 24 bands in order to have a small and controlled experimental setting. First we tested the knowledge-based approach. Due to the fact that we do not have knowledge resources designed for creativity analysis, we exploited Wikipedia, extracting the length of the page (in characters) for each band as a measure of social acceptance, based on the idea that the importance of each band corresponds to the quantity of text stored in the collective memory, represented by Wikipedia. In this experiment we tested how social acceptance can be predicted from the style feature space, made of words of album and song titles. To predict social acceptance scores, we used a regression algorithm based on Support Vector Machines (SVMreg) [26]. With this algorithm, we can compute a formula that predicts social acceptance scores from the style feature space and evaluates the error in the prediction. In practice, the algorithm assigns a numerical weight to the words, based on their ability to predict social acceptance score, and compare the prediction with actual social acceptance score values, reporting the error. To evaluate the error, we used root mean squared error (also called root mean standard deviation) that represents the sample standard deviation of the differences between predicted values and observed values.

We normalized the social acceptance scores and the occurrences of words, in order to obtain values between 0 and 1. Then we applied the "BestFirst" feature selection algorithm [27], that evaluates the correlations between words and social acceptance scores. This feature selection algorithm is a dimensionality reduction technique, that explores the style feature space and considers the individual predictive ability of each word and its degree of redundancy, selecting only the most relevant ones. In the experiment we tested a setting with all 1000 features and a dimensionality reduced space with 26 features. We split the data in two parts: 66 % of the data for the training phase and and 33 % for testing and evaluation. We compared our results to a baseline obtained predicting the acceptance scores with the average value computed on the training dataset. The results, obtained with the majority baseline and with the regression algorithm (with and without feature selection) are reported in Table 1.

These results reveal that, despite the baseline obtained a very low error, the regression algorithm can reduce it a lot when we apply feature selection. This experiment reveals that, given words in song titles, we can automatically predict the social acceptance with an error of 0.087. But the most interesting thing is that with feature selection we can find the most relevant words and their weights in the predicton of acceptance scores, reported below and divided between positive and negative.

Table 1 Results of the experiments

Algorithm	Features	RMSE
Baseline	1000	0.192
SVMreg	1000	0.211
Baseline	26	0.192
SVMreg	26	0.087

We used a regression based on support vector machines (SVMreg) as classification algorithm, and we tested two different settings: with and without feature selection. As feature selection we used a correlation-based algorithm that evaluates the worth each attribute considering its individual predictive ability along with its degree of redundancy between the selected ones. Feature selection reduced the number of features from 1000 to 26

```
+        0.0396 * Along
+        0.0396 * Comin
+        0.0353 * Coming
+        0.0348 * Got
+        0.0479 * Harder
+        0.0451 * Hey
+        0.0806 * Let
+        0.0182 * Lose
+        0.0343 * Luck
+        0.0928 * Me
+        0.1067 * Must
+        0.0635 * New
+        0.0519 * Sand
+        0.0962 * Yes
+        0.1272 * Yet
+        0.099  * Yourself
-        0.0653 * Death
-        0.0038 * Here
-        0.0295 * Flaming
-        0.0165 * Tight
-        0.0129 * Two
-        0.0743 * Makes
-        0.0872 * Weirdness
```

In the second stage we tested the agnostic approach: We use a simple K-means clustering algorithm with euclidean distance, evaluating the clusters on the band's names. In order to assess the creativity/uniqueness of an author, we can set the number of clusters we want as the same number of instances we have, and test whether authors are clustered alone or not. The authors that are not clustered alone are less unique than others. In our case, as reported in Fig. 3, all the authors have their own cluster, meaning that they are all unique. If we want to understand which one is the most distant from the centroid, we have to set 2 clusters. In our case it resulted that Joan Baez is the most unique author in terms of words used in song titles and lyrics, with respect to the other bands.

4 Conclusions

In this paper, we presented a computational framework for HCA inspired by computational linguistics and NLP. This framework has two approaches, one agnostic, based on clustering, and one knowlegde-based, based on supervised learning and feature selection. While in the agnostic approach can reveal the uniqueness of

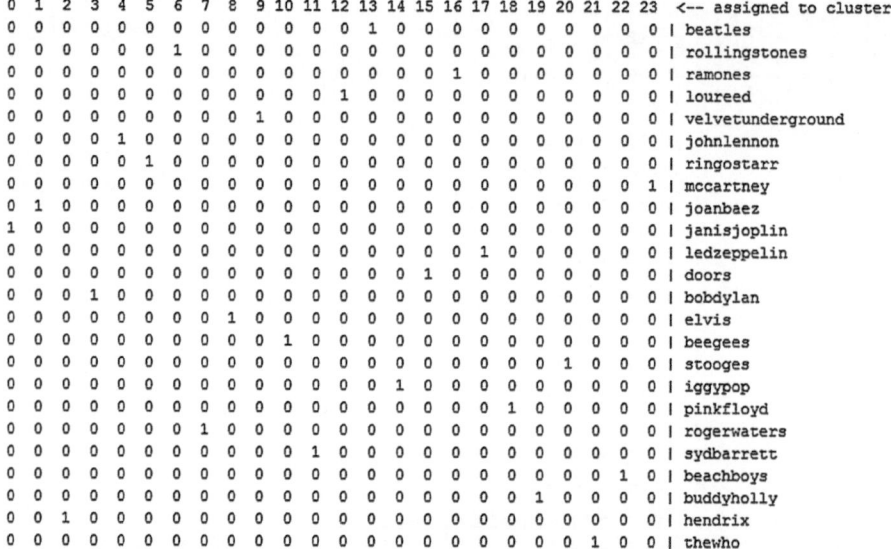

Fig. 3 Results of the agnostic approach to HCA. We set the number of clusters we want as the same number of instances we have, and test whether authors are clustered alone or not. In this case yes

authors from their work, the knowledge-based approach can be exploited to extract the most accepted features of works and to predict social acceptance. In both the approaches, it is required a great effort to define symbols to represent meaningful cues in creativity and style.

In the future, we would like to do new experiments of HCA on a very large scale.

References

1. Csikszentmihalyi, M.: Society, Culture, and Person: A Systems View of Creativity. Cambridge University Press, Cambridge (1988)
2. Csikszentmihalyi, M.: Flow and the Psychology of Discovery and Invention. HarperPerennial, New York (1997)
3. Kraft, U.: Unleashing creativity. Sci. Am. Mind **16**(1), 16–23 (2005)
4. Guilford, J.P.: Intelligence, Creativity, and Their Educational Implications. RR Knapp, San Diego (1968)
5. Torrance, E.P.: The nature of creativity as manifest in its testing. The Nature of Creativity pp. 43–75 (1988)
6. Piffer, D.: Can creativity be measured? an attempt to clarify the notion of creativity and general directions for future research. Think. Skills Creat. **7**(3), 258–264 (2012). doi:10.1016/j.tsc.2012.04.009
7. Kaufman, J.C., Lee, J., Baer, J., Lee, S.: Captions, consistency, creativity, and the consensual assessment technique: new evidence of reliability. Think. Skills Creat. **2**(2), 96–106 (2007)

8. Carson, S.H., Peterson, J.B., Higgins, D.M.: Reliability, validity, and factor structure of the creative achievement questionnaire. Creat. Res. J. **17**(1), 37–50 (2005)
9. Liu, Y.T.: Creativity or novelty?: cognitive-computational versus social-cultural. Des. Stud. **21**(3), 261–276 (2000)
10. Saunders, R., Gero, J.S.: Artificial creativity: A synthetic approach to the study of creative behaviour. Computational and Cognitive Models of Creative Design V, Key Centre of Design Computing and Cognition, pp. 113–139. University of Sydney, Sydney (2001)
11. Wiggins, G.: Categorising creative systems. In: Proceedings of Third (IJCAI) Workshop on Creative Systems: Approaches to Creativity in Artificial Intelligence and Cognitive Science. Citeseer (2003)
12. Barbieri, G., Pachet, F., Roy, P., Degli Esposti, M.: Markov constraints for generating lyrics with style. In: ECAI, pp. 115–120 (2012)
13. Runco, M.A., Nemiro, J., Walberg, H.J.: Personal explicit theories of creativity. J. Creat. Behav. **32**(1), 1–17 (1998). doi:10.1002/j.2162-6057.1998.tb00803.x
14. Boden, M.A.: Creativity: a framework for research. Behav. Brain Sci. **17**(03), 558–570 (1994)
15. Pribram, K.H.: Brain and the creative act. Encycl. Creat. **2**, 213–217 (1999)
16. Mainemelis, C.: Stealing fire: creative deviance in the evolution of new ideas. Acad. Manag. Rev. **35**(4), 558–578 (2010)
17. Runco, M.A.: Creativity: Theories and Themes: Research, Development, and Practice. Academic Press (2010)
18. Averill, J.R.: Creativity in the domain of emotion. Handbook of Cognition and Emotion, pp. 765–782 (1999)
19. Grira, N., Crucianu, M., Boujemaa, N.: Unsupervised and semi-supervised clustering: a brief survey. A Review of Machine Learning Techniques for Processing Multimedia Content, p. 11 (2004)
20. Jain, A.K.: Data clustering: 50 years beyond k-means. Pattern Recognit. Lett. **31**(8), 651–666 (2010)
21. Kotsiantis, S.B.: Supervised machine learning: a review of classification techniques. Informatica (03505596) **31**(3) (2007)
22. Molina, L.C., Belanche, L., Nebot, À.: Feature selection algorithms: a survey and experimental evaluation. In: IEEE International Conference on Data Mining, pp. 306–313. IEEE (2002)
23. Salappa, A., Doumpos, M., Zopounidis, C.: Feature selection algorithms in classification problems: an experimental evaluation. Optim. Methods Softw. **22**(1), 199–212 (2007)
24. Mihalcea, R., Strapparava, C.: Lyrics, music, and emotions. In: Proceedings of the 2012 Joint Conference on Empirical Methods in Natural Language Processing and Computational Natural Language Learning, pp. 590–599. Association for Computational Linguistics (2012)
25. Witten, I.H., Frank, E., Hall, M.A.: Data Mining: Practical Machine Learning Tools and Techniques, 3rd edn. Morgan Kaufmann, Burlington, MA (2011)
26. Shevade, S.K., Keerthi, S.S., Bhattacharyya, C., Murthy, K.R.K.: Improvements to the smo algorithm for svm regression. IEEE Trans. Neural Netw. **11**(5), 1188–1193 (2000)
27. Hall, M.A., Smith, L.A.: Practical Feature Subset Selection for Machine Learning. Springer, Berlin(1998)

Meaning and Creativity in Language

Luc Steels

Abstract The statistical/probabilistic approach has lead to important breakthroughs in language processing and to an impressive impact on real-world applications. This paper examines two issues which have not yet been adequately handled: meaning and creativity. To integrate them, and thus further advance the state of the art, requires a new approach to meaning and a meta-level architecture capable of creative insight into problem solving. I illustrate these ideas using ongoing experiments of language games with autonomous robots.

1 Introduction

The probabilistic approach to language assumes that lexical and grammatical constraints and the grammatical categorisations, which are needed to apply these constraints are probabilistic rather than strictly yes/no. The statistical/probabilistic approach proposes that the required probabilities are obtained through statistical inference over large corpora [7]. Very concrete operational mechanisms to implement this approach have now been worked out in great detail and used as the basis of a broad spectrum of applications.

Today we live in a period where the success of the statistical/probabilistic approach to language is at its peak. This success is well deserved. From a theoretical point of view, there are two benefits:

- The probabilistic approach handles better the obvious flexibility that human users bring to bear on the concrete use of language. The majority of sentences in spoken discourse are fragmented and incomplete, often ungrammatical, and prone to error. A probabilistic approach allows the relaxation of rigid constraints so that a plausible parse and interpretation might still be found.
- Manually coming up with a sufficient set of constraints for a broad coverage of a language has turned out to be a very hard task indeed. For example, no

L. Steels (✉)
Institute for Advanced Studies (ICREA), IBE (UPF-CSIC), Barcelona, USA
e-mail: steels@arti.vub.ac.be

© Springer International Publishing Switzerland 2016
M. Degli Esposti et al. (eds.), *Creativity and Universality in Language*,
Lecture Notes in Morphogenesis, DOI 10.1007/978-3-319-24403-7_13

generative grammar of any language exists that has enough empirical coverage to automatically parse successfully the kind of sentences produced by speakers and encountered by listeners in normal discourse, despite an army of generative linguists working on this. Statistical inference can help to solve this problem, If sufficiently large corpora are available, which is the case thanks to the massive availability of text and speech collected through the Internet, then the acquisition of language models can proceed largely automatically.

It is therefore no surprise that the statistical/probabilistic approach is dominating computational linguistics and its applications at the moment. Systems are not only validated in open competitions but also in their daily use by billions of people. This point was already made five years go by Peter Norvig [9] and is still valid today: all major search engines use statistical/probabilistic grammars, and so do all major speech recognition systems (often relying on probabilistic hidden Markov models). Machine translation systems in practical use (such as Google Translate) and question answering systems (such as IBM's Watson) all rely heavily on statistical and probabilistic techniques. When we look 'under the hood' of natural language processing systems, we see statistical/probabilistic models dominating as well: for word sense disambiguation, coreference resolution, part of speech tagging, parsing, etc. The recent excitement around deep learning only further adds to this general wave of theoretical advances and applications.

Although the success of statistical/probabilistic models is undeniable, there are nevertheless still major unresolved issues which show up as incomprehensible 'glitches' to end-users. For example, although IBM's Watson excelled in Jeopardy, it made the strange mistake of classifying Toronto in the United States. The answers given by search engines are often remarkably appropriate but always approximate and the user has to select in how far they are correct. Thus if you type in Google the query 'politicians who are not belonging to the democratic party' you are directed to a number of sites: about the democratic party, the notion of political affiliation, independent politicians, a site about a certain Will Rogers who has said "I am not a member of any organized party—I am a Democrat.", etc. The reasons for these results is that they have in their text the main keywords of the sentence: politician, belonging, not, democratic, party. However, Google does not make a deep syntactic analysis or grasp the meaning of the query so that the answers are good guesses on which websites information might be found, but they are not necessarily to the point. For example, the proper scope of the word "not" is not handled at all.

Glitches become even more obvious when language needs to be produced, such as in automatic translation. Here are some examples from Google Translate for an English to German translation (downloaded December 2015):

1. a. *Eng:* He does not help the man
    b. *Germ:* Er nicht dem Mann zu helfen
    c. *Lit:* He not the man (dative) to help
    d. *Correct:* Er hilft dem Mann nicht

a. *Eng:* He does not help the woman
b. *Germ:* Er nicht die Frau, die helfen
c. *Lit:* He not the woman (nominative), the helping
d. *Correct:* Er hilft der Frau nicht

Clearly the Google translation messes up German grammar. The position of the negation "nicht" is incorrect in (1) and (2), the case is wrong in (2), the verbal form (finite main verb) is not recognized and translated wrongly in both (1) and (2). We clearly see here the limits of the approach.

Because of these glitches, users of statistically based AI systems often have the impression that they are dealing with an 'idiot savant', who is sometimes remarkably intelligent, responding in the right way even to a difficult request, but at other times clueless. Improvements can be expected with more data so that 'ready-made solutions' are in the corpus and can therefore be picked up by the statistical algorithms, but more data can also make matters worse because specific cases which are rare but nevertheless correct in specific circumstances, can get snowed under. What exactly is missing and how can we move forward?

2 Where Is Meaning?

Suppose we have data about the behavior of a person, let us call her Emilia: each weekday morning she opens the door of her house at roughly the same time in the morning and steps out. She goes to a particular car, opens the car, turns the ignition key, gets out of the parking place, and drives to a particular location in the city. A statistical learning algorithm could easily build a probabilistic model of the sequence of these behaviors and therefore predict what is going to happen next. Some days Emilia may first drive to a pharmacist, or drive somewhere else before going to work, or get out of the house later. So the probabilistic model could be refined and adapted as routines change. However, what the model does not grasp at all is the causal relations and intentions behind these behaviors. Emilia goes to the car *because she wants to drive to work*. She opens the door *in order to get in*. She turns the ignition key *to start the engine*, and so on. Integrating these causal relations requires an additional layer of goals and subgoals and a planning system that can make and recognize plans.

Statistical natural language processing is similar to such a statistical action predictor. It is observing words or syntactic structures without integrating causes and then trying to predict the next word in a sequence. But a human speaker does not utter the word "with" after "the man ..." because here are a lot of occurrences of the sequence "the man with ..." in her corpus of earlier utterances (even assuming that humans store such a corpus), but rather because she is trying to refer to a particular person and uses a description which involves an appropriate feature of the man, as in "the man with the umbrella". Similarly, a statistical translation system learns the most common n-gram (sequence of n words) in a source language that is paired most often in the corpus with an n-gram in the target language, but it does not at all take into account why these words have been chosen by the human translator as the best translation.

The same holds for grammar. Grammar is meaningful and so there is a reason why a particular grammatical construction is used. Grammar either expresses meanings which go beyond the meanings of individual words. For example, in "Paul painted the chair black" the grammatical pattern "NP1 Verb NP2 Adj" expresses that NP2 (the chair) results in a state described by Adj (being black) after the action Verb (painted) is carried out by NP1 (Paul). Or grammar helps to avoid combinatorial complexity in parsing. For example, the phrase "Miquel has never played rugby before is apparent" is a garden-path sentence that leads us astray because we expect the first clause ("Miquel has never played rugby") to be the main clause. This wrong path is avoided by adding a grammatical function word "that" as in: "That he has never played rugby before is apparent".

All this suggests that we cannot ignore the issue of meaning when we want to go beyond current levels of performance. The expression or reconstruction of meaning drives human language and hence the ultimate goal of computational linguistics is the proper comprehension and production of language based on meaning.

However, meaning is the achilles heel of statistical natural language processing because meaning is not observable and hence it is not possible or easy to assemble corpus data as input for a learning system. More recent statistical language processing experiments therefore try to capture meaning in an indirect way.

A typical example is found in Bottou [4]. A set of words is represented with an n-dimensional vector $W : words \rightarrow \mathbb{R}^n$, initialized randomly for each word. The values of the vectors function as the weights of a neural network that has to decide whether a presented word occurs in a corpus or not. Learning takes place by taking sentences from the corpus as positive and generating sentences with some random change (for example a random word inserted) as negative.

Surprisingly, the word vectors cluster according to syntactic and semantic criteria, for example, putting names of countries, names of consumer technology devices, names of colors, etc. together. Moreover vector offset methods based on cosine distance capture syntactic and semantic relationships (e.g., between singular and plural forms of words (year/years) or between male and female nouns (king/queen)) [8]. These results show that statistical methods are a very powerful road towards learning linguistically relevant skills, not just at the level of words but also at syntactic and semantic levels. They also show that statistical learning has a role to play also at the level of semantics although these systems are piggybacking on human-produced purposeful language and still do not address meaning directly.

3 Where Is Creativity?

An advantage of probabilistic language processing is that it copes to some extent with the tremendous variation found in normal language use, particularly in spoken language. Each speaker has his/her own idiolect and lexicons and grammars are changing all the time, which implies that listeners must be prepared to deal with sentences that are not conformed to their own systems. Probabilistic decision-making allows that categorical decisions, for example whether a word can fit in a particular grammatical construction or whether a grammatical construction can apply, no longer needs to be made in a yes/no manner. This helps to deal with performance deviations due to errors in pronunciation, wrong word choice, inappropriate use of grammatical constructions, or mistakes in conceptualization.

However, there are also many deviations of standard language use that are intentional. Language systems are not static. Speakers have the right and are able to extend at each point their language systems in order to express meanings that have never been expressed before or might be expressed in a more original forceful way. Listeners have to follow suit, guess the novel meaning of a word or stretch and expand existing grammatical constructions, and then possibly store these innovations as part of their own language system which thus starts to propagate in the population.

Creative language use is much more common than one may assume and it happens at all levels of language. Here is an example:

1. *He was the Massachusetts Board of Health's go-to researcher for all kinds of sanitation queries.* (source: MIT Technology review, Dec. 2013).
2. *Their opponents know these go-to players will be getting the ball.* (source: Forbes, Oct 2013)

The word 'go-to', seen in Example 1, has been formed by a contraction of the verbal form "go to" to a single word which then became coerced in this context to behave as a nominal modifier. It can only be applied to persons and means somebody who excels and therefore is seen as an authority. Although you may never have seen Example 1 before, it is immediately comprehensible, and we can easily infer the syntactic moves (contraction and coercion) that the speaker made. In Example 2, the use of 'go-to' is mapped to the domain of sports. It now refers to a person that is seen as the best and therefore the one who gets the good passes that may result in a goal. Incidentally, Google Translate translates "a go-to person" in French as "le feu à la personne" (literally: the fire to the person) which is puzzling and incomprehensible. When "le feu à la personne" is translated back into English we mysteriously get "the fire the person."

Here are some more examples:

1. *It just sounds a bit namby-pamby, does not it? A bit comfort-blanket.* (Source: Guardian commentary August 2014)
2. *Anyway—that's my five-minutes-a-day-just-in-case Guardian time up, now I can concentrate on the media that is relevant to the debate.* (Source: Guardian commentary August 2014)

"Namby-pamby" in Example 1 is in fact an old eighteenth century word for affected, weak speech. But (at least for me) it was a new word that nevertheless conveyed through its sound the intended meaning of a fuzzy wooly statement (by a politician). The word "comfort-blanket" is probably entirely novel as a way to describe a statement, but it nicely conveys the same meaning as "namby-pamby" through a metaphor.

The noun phrase "my five-minutes-a-day-just-in-case Guardian time up" in Example 2 builds a new noun compound "Guardian time up" from a sentence "my time is up", to mean the time I have available for reading the Guardian newspaper. It is an entirely novel creative construction, which illustrates the remarkable capacity of speakers to come up with very compact ways of saying something. The phrase "five-minutes-a-day-just-in-case" is a typical example of a current trend in English to push longer and longer adjectival phrases into the nominal phrase. It basically says 'five minutes a day just in case there is something interesting'. The phrase is syntactically speaking a noun phrase but coerced to function as an adjective.

These examples come from commentaries to newspaper articles and illustrate better common ordinary language use than polished written text or sentences invented by armchair linguists. They illustrate well the extraordinary creativity of language usage—which goes entirely beyond the grasp of statistical language processing techniques.

The question is then in which direction we should go in order to bring meaning and creativity into language studies and language processing applications, without losing the powerful results obtained through statistical/probabilistic models. Clearly we need to bring in techniques from earlier work in 'symbolic AI': sophisticated knowledge representation, perceptual grounding, common sense inference, planning, richer grammar. Quite a bit of research has already gone in this direction with so called 'hybrid systems' that combine, for example, rule-based, hierarchical grammars (e.g., dependency grammars) with statistical/probabilistic models (such as hidden Markov models) [6]. I will introduce some more ideas to pursue this further in the next sections.

4 Bringing in Meaning

The first step is to see language interaction not as a kind of Shannon transmission process with a sender that codes a message and a receiver that decodes it, but as a Wittgensteinian language game [10]. A natural language game is a routinized interaction in a joint context. There is some communicative intention from the side of the speaker and language is only one aspect of the interaction. A language game is based on inferential coding which means that not all information is in the message itself but some of it has to be inferred indirectly from the shared context, common sense knowledge, and prior discourse. There is an interlacing of comprehension and production as well as constant feedback and repair to make sure understanding is achieved.

The language game paradigm is of interest here because it allows us to bring in the notion of (grounded) meaning. Meanings are distinctions that are relevant in the interaction between an agent and the environment, which includes other agents. For example, the distinction between red and green is relevant for traffic lights because when we ignore it or interpret it wrongly we may get killed. The internal representation of a distinction is usually called a category. Categories and hence meanings are grounded if a systematic link is made to sensory-motor experience. Sensory-based categories, e.g., 'red' versus 'green', are perceivable through sensing, signal processing, segmentation, and pattern recognition. Action-based categories, such as 'stop' versus 'go', must be transformable into concrete behaviors that have an effect in the world.

There is now a substantial literature on language games in software simulations and even with autonomous robots (see Fig. 1 from [3]). A language game experiment can be set up in which all the necessary components are completely programmed in. For example, the robots are initialized with the necessary pattern recognition and classifier systems to handle a particular set of categories and a lexicon and grammar to communicate with them. However, given the complexity and adaptivity of language and meaning it is more interesting to acquire both the categories and conceptualizations as well as the means of expressing them through learning strategies. Many of these use the same techniques as we find in statistical machine learning and they also acquire probabilistic grammars and lexicons. The main advantage of using language games in contrast to corpora is that meanings are available (although not necessarily shared) because the partners in the language games both have access to the shared context and both are acquiring the necessary categories to deal with the environmental challenges posed by the game [5].

This is not the place to go into further detail on the kinds of strategies that have been shown to lead a population to self-organized shared conceptual and linguistic inventories. Figure 2 from [2] shows just one example from experiments in the emergence of grammatical agreement (such as agreement between subject and verb for

Fig. 1 Language games between autonomous robots. The robots play a color naming game in which one robot chooses a color chip, names the color, and the other robot has to point to the chip which best represents this color. Robots start without any color categories or color words and build up and negotiate a shared color ontology and vocabulary

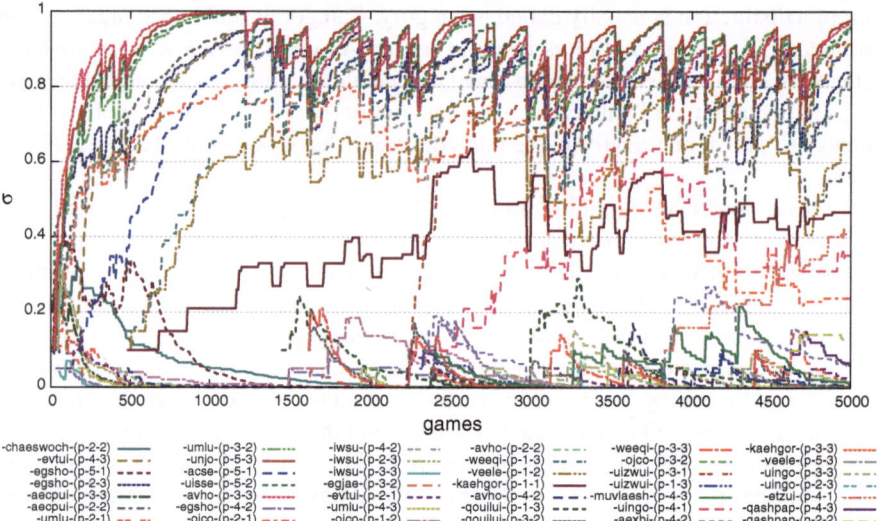

Fig. 2 Emergence of grammar of agreement. These data are from an experiment where a population develops a system of markers to signal that different constituents (e.g., an article, an adjective and a noun) belong to the same parent phrase. The markers are recruited from existing words that take on a new function. There is a general winner-take-all dynamics but, because of a flux in the population, new markers may appear that potentially but rarely overtake existing markers

person and number in German, or article, adjective, and noun for gender and number in French.) Many more examples are provided in reference [11]. I also refer to [1] as an example of the statistical physics approach towards the study of the semiotic dynamics that arises in these language games and experiments with human, and [5] as an example of psycholinguistic experiments showing empirical evidence for the strategies that have been shown to be effective in computer simulations and robotic experiments.

5 Creativity Through Insight Problem Solving

As argued earlier, creativity is more than making random changes. Language creativity is purposeful, seeking to express new meaning by reappropriating existing language material, or finding more expressive ways for existing forms of expressions. I therefore make a second suggestion: We have to approach language processing as a problem solving process based on a two-tier system. The first layer is concerned with routine language production or comprehension. It is highly automated and very fast. Many ready-made patterns are recognizable and usable in an instant, otherwise we cannot explain why we can speak so fast or understand fast speech.

But there has to be a second layer, a meta-layer. It runs diagnostics, monitoring routine language usage, and comes in action when an anomaly appears, for example, the listener hears an unknown word, the lexical category of a word does not fit with the expectations created by a particular grammatical construction, the listener gets in a cul-de-sac, i.e., a garden-path sentence from which she has to backtrack to make sense of the utterance. The speaker may also get in trouble. There is a concept for which no word exists yet, a particular way of phrasing was chosen in which some of the meaning she wants to express cannot be made to fit, she may not remember a word, or she may try to speak or write more compactly as is normally allowed in the language.

Once an anomaly is detected, a repair action comes into action. The repair action tries to find a possible solution to the problem and, if this solution lead to a successful interaction, it gets stored for future use. A novel solution might be a one-time event, for example because the listener did not get it or because both of them did not store the outcome, but it may also start to propagate in the population and possibly become the new norm.

A concrete example where this architecture has been used is an experiment in the emergence of phrase structure grammar, reported fully in [12] (See Fig. 3). This experiment does not use a statistical learning approach with a large corpus, but sets

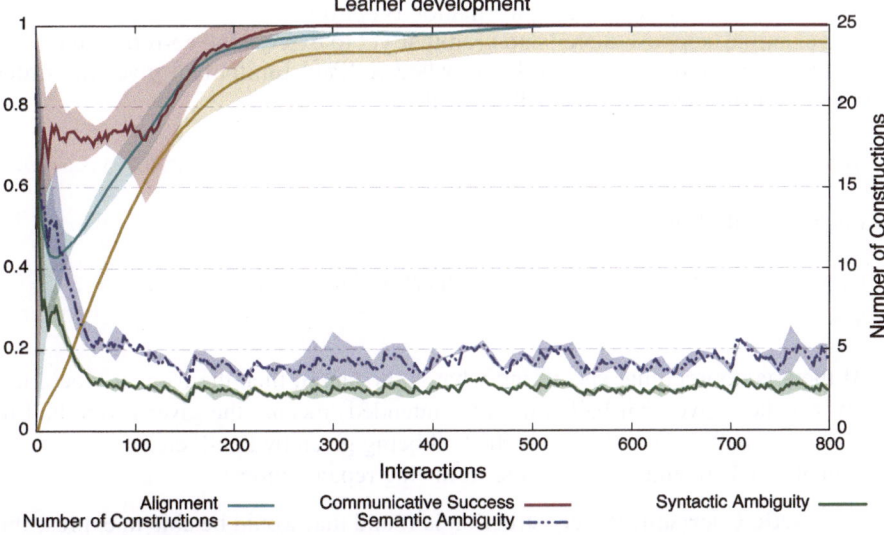

Fig. 3 Emergence of phrase structure grammar. These data are from an experiment where a population of 5 agents plays language games to refer to objects in the world and develops a phrase structure grammar using creative meta-level operators. The graphs (for 10 run with average and standard deviation) show for a series of 800 language games the *alignment*, which measures whether the hearer would express the same meaning using the same utterance as the speaker, *communicative success*, which measures whether the hearer was able to identify the topic without speaker feedback, *syntactic ambiguity*, which measures the number of extra hypotheses that grammatical constructions generated during processing, *number of constructions*, which is the average number of constructions for all agents—it is reaching 25, and finally the *semantic ambiguity*, which measures the number of times the world model of the situation is used to generate or block hypotheses, divided by the number of variables introduced by the lexicon for the utterance

up agents playing language games about situations in the world. The speaker refers to an object (called the topic) by describing a situation and the listener has to figure out from this description which object is intended. The game is similar to the naming game used in many other language game experiments (see Fig. 1) but agents can now use relations with multiple arguments (instead of just unary predicates) and that requires them to come up with some form of syntax. Agents learn in an incremental fashion using a situated embodied interaction and also expand their construction inventories.

In this experiment, the agents use a description calculus for meanings. The description calculus is based on the predicate calculus but has an operator to signal which one of the arguments of a set of predications is the 'target' τ, i.e., the object being referred to. For example. $\tau(x)ball(x), big(x)$ is expressed as 'the (or a) big ball'. $\tau(x)(give(e, x, y, z), \tau(a)ball(a), \tau Paul(b), a = y, b = z)$, where e refers to a give-event, x to its agent, y to its object and z to its recipient, is expressed as 'the giv-er of the ball to Paul'. $\tau(x)(ball(x), give(e, x, y, z)), \tau Paul(a), a = z)$, as 'the ball given to Paul'. Words in the lexicon convey the predications and grammar must convey (i) which variables in these predications are co-referential and (ii) which variables are targets. The diagnostic run by the speaker therefore tests for a particular utterance whether this information is conveyed.

For example, suppose there is no grammar yet to express the co-referential relations then the listener would not know whether 'Paul Emilia ball give' (no order intended) means Paul gives the ball to Emilia:

$$\tau(e)(give(e, x, y, z), \tau(a)Paul(a), \tau(b)ball(b), \tau(c)Emilia(c), a = x, b = y, c = z),$$

or Emilia gives the ball to Paul:

$$\tau(e)(give(e, x, y, z), \tau(a)Emilia(a), \tau(b)ball(b), \tau(c)Paul(c), a = x, b = y, c = z)$$

If there is no grammar specifying what is the target, then the listener would not know whether 'give paul ball' (no order intended) means 'the giver of the ball to Paul' or 'the ball given to Paul' or 'the ball being given by Paul', etc.

Agents are here endowed with the following repair actions:

1. **Syntactic Coercion:** If a construction is found that would be able to express the co-reference relations and targets from a semantic point of view but the lexical categories of the words being used do not fit with the slots in this construction, then the words can be coerced to fit these roles.
2. **Semantic extension:** If a construction is found which expresses the right co-reference relations and targets and the syntactic constraints match but a required semantic category does not fit, then the construction can be adapted to accept the semantic extension.
3. **Novel basic construction:** If no construction is found, then a new one is built. It has slots for all the words that introduce the predications and imposes a particular syntactic order on them. The lexical categories of these words are the ones already

used with the words or else new lexical categories are created and assigned to the words. The construction also introduces a hierarchical phrasal unit for the target which inherits its semantic properties from the target slot.

4. **Novel Hierarchical construction:** It is possible that the slots contain not a single word but a group of words that was already combined into a hierarchical unit. Then a new construction can be built in the same way as in (3) except that the syntactic constraint on the slot now specifies the phrasal category of the hierarchical unit filling the slot.

We refer to the literature for a more elaborate exposition of these repair strategies. The results in Fig. 3 show that these diagnostics and repairs are enough to see the emergence of a phrase structure grammar. The grammar even exhibits recursion, progressively allowing phrases such as: "the pyramid on top of the block sitting on the table standing on the floor inside the room". We see that agents reach steady communicative success, meaning that they identify targets and co-reference relations, we see that syntactic and semantic ambiguity get damped, and agents align their grammatical usage.

6   Conclusions

This paper argued that the statistical/probabilistic approach has made great strides recently and is now undisputed as an important contribution to natural language processing. However, there are two points of weakness: meaning and creativity. I argued that meaning can be tackled by shifting from corpus-based learning to learning through language-games. A corpus contains mostly only utterances without a representation or enough contextual information of the communicative goals and semantic structures that underlie these utterances. Language games introduce meaning because speaker and listener have shared goals and a shared context and meanings are generated and can be deduced. I furthermore argued that creativity can be brought in by introducing a two-tier architecture with routine language processing interlaced with meta-level insight problem solving to handle the intentional deviations that speakers introduce in order to adapt and expand their language systems. Experiments with language games already reported in the literature show the power of the approach but more researchers need to get involved to scale up these experiments to achieve the full potential impact.

References

1. Baronchelli, A., Felici, M., Loreto, V., Caglioti, E., Steels, L.: Sharp transition towards shared vocabularies in multi-agent systems. J. Stat. Mech. P06014 (2006)
2. Beuls, K., Steels, L.: Agent-based models of strategies for the emergence and evolution of grammatical agreement. PLoS ONE **8**, e58960 (2012)

3. Bleys, J., Loetzsch, M., Spranger, M., Steels, L.: The grounded color naming game. In: Proceedings of the 18th IEEE International Symposium on Robot and Human Interactive Communication, Roman (2009)
4. Bottou, L.: From machine learning to machine reasoning. Mach. Learn. **94**, 133–149 (2014)
5. Centola, D., Baronchelli, A.: The spontaneous emergence of conventions: An experimental study of cultural evolution. Proc. Natl. Acad. Sci. USA **112**(2), 1998–1994 (2015)
6. Hirschberg, J., Manning, C.: Advances in natural language processing. Science **349**(6), 261–266 (2015)
7. Manning, C.: *Probabilistic Syntax* In: Bod, R., Hay, J., Jannedy, S. (eds.) Probabilistic Linguistics, 289–341. MIT Press (2003)
8. Mikolov, T., Yih, W., Zweig, G.: Linguistic regularities in continuous space word representations. In: Proceedings of NAACL-HLT, pp. 746–751 (2013)
9. Norvig, P.: On Chomsky and the Two Cultures of Statistical Learning. http://norvig.com/chomsky.html (2012)
10. Steels, L.: Language games for autonomous robots. IEEE Intell. Syst. **2001**, 17–22 (2001)
11. Steels, L. (ed.): Experiments in cultural language evolution. John Benjamins, Amsterdam (2012)
12. Steels, L., Garcia-Casademont, E.: How to play the syntax game. In: Andrews, P. et al. (eds.) Proceedings of the European Conference on Artificial Life 2015. The MIT Press, Cambridge. pp. 479–486 (2015)